NERO'S VIOLIN

The art of conviction in a divided world

Johan van den Berg

First published in South Africa in 2015 by EE Publishers.

Republished 2016 by Horse of Troy.

Print ISBN: 978-0-620-67821-6
E-book ISBN: 978-0-620-67822-3

Front cover image: “Nero’s Violin”, by Mary-Jane Persson, 2014 (oil on canvas)
Back cover image and additional colour images of Nero with violin: Jade Holing
Pencil art: Kerri-Ann Venter
Additional artwork: Alex van Houwelingen
Language editor: Bridget Theron-Bushell
Author’s technical book advisor: Tracey McDonald
Final copy editing: Elizabeth Donaldson
Design and layout: Fire and Lion

CONTENTS

ACKNOWLEDGEMENTS

I am thankful to a range of people who assisted with the book:

The following people read early drafts and/or took part in peer review, contributing significantly to improvements and refinements that I hope will be evident in the final version (alphabetically by surname): Heidi Botes, Mareike Britten, Zenda Cornelissen, Soren David, Sol Luca de Tena, Joke Debaere, Amanda Dinan, Louise du Toit, Leole Francis, Andrew Gilder, Leonie Joubert, Francois le Roux, Alan Lightman, Ruth Mhlanga, Stefanus J. Minnaar, Martin Ogle, Vanessa Otto-Mentz, S. Pamela Paquin, Jeunesse Park, Karin Pautz, Johan Roos, Steve Sawyer, Frank Spencer, Bridget M. Theron-Bushell, Toy van den Berg.

The following people enriched my thinking and understanding of the subject matter by dedicating time to discussion (alphabetically): James Cameron, , Marian Goodman, Ralph Hamann, Christine Lins, Julia Martin, Liz McDaid, Naomi Oreskes, David Peat, Gary Snyder, Nicky Stilwell, Clem Sunter, Ciska Terblanche, Emily Wilson, Holle Wlokas and Kim Wright.

The intrepid Marilize Stoltz assisted most capably with a very wide range of challenges.

Prologue

Imagine we're on a yacht in the middle of the ocean, six of us. We can all sail – that's how we know each other. A single hull yacht for recreational use. Engine for an emergency, fuel for fifty miles. We hired the boat with a captain, then carried on without him when he fell ill and had to be dropped off the previous day, with all his fancy, electronic weather equipment.

No need for a final decision-maker with power over life and death, we're all civilized and reasonable people.

On days when the wind blows, we each have a job to do. We become a team. Indistinguishable from the yacht. The yacht and the six of us become a single, well-oiled machine with astounding abilities to propel itself, steer itself, navigate, repel water, stay upright, change direction, change speed. We shout friendly directives at each other to co-ordinate our efforts.

But today there is nothing. Not a breath. Some swell yes, moving the hull this way and that. And humidity. But quiet. There are no sails to attend, no steering to be done, no navigation to do. Until further notice, a day of leisure.

You and I are taking turns inside, at the radio. Two hours at a time. We're a little too far away from land and reception comes and goes. Then it improves, and a report comes in. It talks about light rain, but nothing else. No wind or anything unusual. I'm there. I look out at you and the others, wondering if I should update you. Sam is enjoying the sun on his tanned body. Erin is reading, Lorenzo playing with an i-pad. I decide not to say anything, not to spoil the fun. If the rain comes, we can all move inside. Erin can carry her book down, Sam can put on his shirt all by himself. The little inconveniences that arise from the report being true can be dealt with individually, without even talking to each other. Even if Sam decides to stay out and get wet, Erin can move inside. Each can determine an own day and schedule.

Then we change over. You come in to the radio, I go out to sit down and stare across the water, marvelling again at the impossibility that nothing but nothing else can be seen and that even on the horizon, it is unclear where water ends and the air begins.

You're just starting to relax, reading a magazine, when the report comes in – short and truncated, severe and anxious. Cyclone warning. Wind speeds up to a hundred miles an hour, swells up to forty feet, buckets of rain. We have to start preparing immediately, and sail as soon as we can into open seas. Away from the reefs around the island. You turn up the volume and get ready to call us in, to hear it. But the radio cuts out.

Gone. Quiet.

No decent reception of anything else anywhere nearby, just a Russian trawler indecipherable and hardly audible, its proximity unknown and unknowable. You run through the frequencies again, somewhat frantic. Looking for the repeat, the confirmation, wanting to call us to hear it.

But there's nothing.

You walk out then, briskly, like you're about to break into a jog. This is serious, not a time to go slowly, but the yacht is too small to save any time that way.

What words? You just let it come the way it wants to, haltingly, then again too quickly. Somewhat cool, somewhat anxious.

You look at the five faces, instinctively closing in on you, huddling. I'm already convinced. I've just been at the radio, listening for something important. And the others in a split second will be convinced also, when the huddling has brought us really close together. Just then Sam says we must all go in and listen together.

The moment of acceptance is suspended. The huddling is halted, we all stay where we are.

"It's gone..." you say. "The frequency is dead. I heard it and then it was gone..."

Sam looks over his shoulder at the limp sails, the blueness, the brightness, the cascading sunshine.

"Somewhere else ..." he says, and moves back into the sun and onto his towel. "The cyclone's probably many miles away. Nowhere near us."

We look at each other, you and me, and in the instant we understand that we have only a moment to seize the initiative, we also realise it's too late. That there was only a moment ... and that moment was taken up realising it. The others retreat, smiling reassuringly at us and at one another.

And you and I look at each other, speechless. One woman and one man cannot save the yacht. If the others don't believe the report, we might as well disregard it too. We could all sunbathe or read, because the two of us can't do this alone.

This book is about being on that yacht, and being one of the two. About science and stories. Collaboration and competition, love and hatred. About how we convince and align the others, knowing our lives and theirs depend on it.

INTRODUCTION

The story of a beautiful lie: Nero and his violin

Rome: AD 64

It was 18 July of year 64 AD.

Very, very late in the day, but his grandfather Mark Anthony, had he lived, would have been able to prevent the carnage about to be unleashed. Would have been able at this time of crisis to rise with the same firm voice that had proclaimed to all friends, Romans and countrymen that he'd come to bury Caesar – not to praise him. Standing over the still-warm body in the pool of blood. The firm voice that had then gone on not to bury Caesar, but to praise him. Mark Anthony would have been able to rouse the Romans to look for six men with torches, pretending to be drunk, about to set their city ablaze. Would have been able to shout with utter conviction "... I proclaim not that my grandson Nero is a crazed murderer, but that our city is in grave danger..."

And everyone would have thought of him as a crazed murderer and gone hunting for the six men with torches and then come hunting for him.

Nero stretched and suppressed a small yawn. Not that he was sleepy, he was as alert and aroused as he'd only been a few times before. When he became emperor almost a decade earlier, he felt this same rush. The next year, when he had his brother Britannicus murdered, he also felt this giddy elation. Until then it was always his mother Agrippina ordering the deaths, but suddenly she had Britannicus lined up as emperor. The son who would listen and obey and allow his mother to rule. Britannicus had to go, but it felt good to send him on his way. Ultimately powerful. Telling a gladiator, "kill him" – and knowing it would be done.

In later years he started being more specific. He'd say whether he wanted his enemies to suffer. How he wanted them killed. With Britannicus he was just cutting his teeth. Only the result mattered.

Nero brought a goblet to his lips and sipped on his wine. His eyes didn't move from the city, lying below his palace on the Palatine Hill. The sun had set after a clear and hot summer's day, and the last orangey-red trace in the west was about to disappear.

"Wait until after dusk," he'd instructed the six gladiators before they each left in a different direction, Remember that we must be able to say someone else did this. Your lives depend on it, as well as mine. And pretend to be drunk, just in case any of you are caught.Say it was an accident and I will pardon you."

He put down his wine and stretched again. It was true – he wanted the cover of darkness, the longest night of the year, so the rescue efforts that daytime would bring could be furthest away. But more than anything he wanted to be able to see the fire. Since he'd been a boy, he'd loved fire. Fire and hurting animals. Both gave him this rush, this total high, this walking with the gods.

It'd been harder to achieve in the last few years. Implicating his advisers Burrus and Seneca in corruption shortly after Britannicus went to the next world was convenient but did very little to make the blood gush through his veins. Banishing his mother from court was better – after all, nobody expected it. The way she pursed her lips in utter astonishment and indignation and humiliation as she was led away. Conniving harlot.

And yet, she didn't learn her lesson. Couldn't stay out of it. Just couldn't shut up and enjoy the privileges she still had. So he had her shipwrecked. Agrippina the murderess of Caligula. She deserved no less. But she survived. Perhaps spared by gods to teach him a lesson. To make the most of murdering an enemy. To do it directly, leaving only a sliver of doubt about who was responsible. To do it to the knowledge of the senate and the people. So they could fear him. So that he, Nero, could be reminded that he is above everything. That he's acknowledged as dictator in perpetuity and that everyone in the known world lives or dies at his pleasure.

The second time, he had her poisoned. As she died, his delegate told her that Nero wished her well in the next world. That he thanked her for all she'd taught him and that he would continue to apply everything in her absence. Told her as she lay writhing in agony, begging the gods to take her, that her son hoped that her life would be spared as long as possible.

That was a total rush, the highlight of his early years in power. But he'd soon fallen into a melancholy. Having killed her, she was no longer there to hate. He could no longer fantasize about getting rid of her.

Nero missed his mother.

Sending others to join her in the next life failed to lift his melancholy or his heart rate. It was only when he started playing music in public and winning competitions that the intoxication returned. Nailing down everyone's attention not as dictator, but as artist. Not just in Rome but as far afield as Athens, where the classical world immediately recognised his genius and gave him standing applause every time.

No sense there of it being below the emperor to perform in public. None of the Roman senators' snide cynicism. Athens wasn't the artistic wasteland that was Rome. The banal hotchpotch of building styles that had sprung up without any master plan or unifying style.

He let his eyes slide over the dwellings below him, looking for the glow and listening for the cries telling him that the fire had started. That old, decrepit and ugly Rome was being ushered into the next world. That at last the wonders of Athens would have space to prosper again, in this place, amongst these seven hills. The fire that would cleanse the entire Republic from vulgarity and make it the proper and palpable continuation of glorious, glorious Athens. The Athens that had herself fought so long and so valiantly against the vulgarity of Troy and utterly destroyed it when the chance finally came.

Here, below him, it wasn't Athens that had been resurrected but Troy that had evolved. Grown like an enormous parasite, despite being an affront to the gods and anyone with a sense of beauty. An ugly city filled with barbarians.

He would bring Athens here in architecture, but also in music. He himself would do what no other emperor had come close to doing: he would play music in public for the masses.

Just then, two cries rang out from different places. One a woman, shrieking inanely from directly below him. The other, a man shouting an urgent warning from his right. On Nero's hilltop, the uniformity of deep shadow and near silence was disturbed as people scurried, shouted, wailed. Torches held up to grasp what was happening created odd shapes of light and shadow. Nero imagined the shadows to be lions in the Coliseum, let loose to finish off the plebs. Then the yellow and orange salvation of the Republic appeared, delivered by his six gladiators. The light intensified as the flames rapidly grew. It seemed the sun was rising again.

The crowds rushed this way and that, running from fire into fire. It was on all sides, spreading from house to nearly adjacent house, engulfing all.

Athens had space between houses, and broad streets. The new Rome would, too.

Nero went inside and donned his stage costume, laid out for him earlier that afternoon. And picked up his violin, carefully tuned as the sun was going down. When he came out soon afterwards, he was delighted to see the fire was twice as large. Twice as bright. The cacophony more intense, less intelligible, more visceral.

Unbelievable. This was his moment. The instant replacement of one order with another. Culture to return and prosper. He raised his violin and ran the bow over the strings. Then he played, starting slowly and softly but soon building to a crescendo. Taking a deep breath to quell his excitement, Nero started singing, his falsetto voice reciting a song well known.

The sacking of Troy.

Rome was burning.

Nero was playing the violin.

Fiddling. [1]

***

The above is a beautiful lie.

Nero never played a violin.[2]

Maybe he started the fire, maybe not. We don't know. Maybe he sang, maybe not. Perhaps he played a cithara, perhaps not. And perhaps he was in Rome when it happened. Or perhaps not.

He definitely never played a violin. The violin was still waiting to be invented.

But who isn't captivated by the image of a crazed emperor playing a violin against the backdrop of a runaway inferno that he'd instigated? His face and the elegant musical instrument silhouetted against the blaze? His high, unstable voice adding a soundtrack to the intense drama?

The image enchants, it captivates, it seduces. It survives all efforts to replace it with the mundane.

In Rome today, they might say "*Si non e vero, e ben trovato.*" It may not be true, but it is still good to hear.

The beautiful lie is stronger than the mere truth.

***

Not all lies are beautiful. Some are devoid of any imagination. They lack the ability to captivate us or seduce us. But they may still affect us.

A gladiator is walking through Rome in the dusk, preparing to set fire to a neighbourhood. Where will he start? The inhabitants of the first house are likely to perish. The neighbours may have time to flee.

He stops and asks a stranger how many people live in his house. He fakes a drunken slur to his speech. Nero has said to do so, in case they get caught. So they can claim it was an accident. "Just my wife and I," says the stranger, hoping the drunken gladiator will ask to stay over for the night and offer payment. In truth, six people live in the house.

Six. Or two. What difference? The lie is ugly. Commonplace. One figure or another.

But soon afterwards the gladiator returns and sets fire to the stranger's house.

***

If a lie is 'ugly' when it is devoid of any ability to capture our imagination, then so is a truth.

Mere truth in this way is 'ugly'. Mere information. Statistics. We have much of that at the moment.

The distance from Berlin to London Heathrow is 840 miles. The remaining flying time is one hour and thirty-five minutes. There is a headwind of 36 miles an hour.

A man rises to address an international conference about the elimination of world hunger. He tells the delegates how many people live on less than two dollars a day. The number is very large. Some of the delegates make notes, a few look at their watches, a few others yawn. Such a large number. What does the face of a multitude made up of billions of individuals look like? Does he have brown eyes? Does she have curly hair? What does he dream of doing with his life if only he had two more dollars a day? What does she think when she sees people driving by in limousines costing as much as all the food she will eat in her entire life? Spread over more than a billion individuals, these potentially beautiful peculiarities disappear.

Truth can be ugly.

To those believing in the cause and working every day to reduce world hunger, ugly truths are important. To them it isn't ugly. They see the subtle differences. They will make notes and draw graphs to plot trends since the previous occasion when the international conference met. They will calculate the rate of change and say in their reports that the rate at which world hunger has increased is slowing down. From this they will draw encouragement to work harder.

But to someone who came to the conference for the first time, wondering if she should spend her scarce spare time and money fighting world hunger, the statistic won't be captivating. The truth that moments before had its hand around her heart (and only needed to close to seize it) will not do so. The heart and its owner will go to another conference about another topic in the hope of being seized by something beautiful. Or go to a music concert and forget about it all.

***

For some centuries now, many scientists have believed that nature speaks for itself. That the truth speaks for itself.

They believe this because this is the way they would like the world to be.

If, in a big city on a cold morning, such a scientist should stop at a pavement coffee vendor enveloped in vapour rising above the icy pavements from the subway below, order a double espresso and walked away moments later to her office with a steaming paper cup, there would be spring in her step at the thought that truth speaks for itself. She would be able to say to herself that her job is just to discover truth. This is usually difficult. But once achieved, her work is done. She need not communicate the truth. Most definitely she does not need to sell the truth. Selling is not what she does. She does science. She gets to her desk with a half-full cup and switches on her PC with a smile on her face.

To these scientists, truth and falsehood lie on a single axis, like this:

truth——————lies

They see their occupation as advancing possible truths towards the edges of this line. From 'hypothesis' somewhere in the middle. No matter which way. Towards the left and certain falsehood, or towards the right and provisional truth. They then see their job as done. Which sometimes it is, and sometimes not.

If such a scientist were to have overheard the discussion between Nero and the six men charged with pretending to be drunk and setting fire to Rome, he may have considered all the evidence ... and after careful consideration might have written on a piece of parchment: "Nero is planning to destroy the city of Rome by fire."

This conclusion is correct. The truth would have been discovered. But if the scientist then filed the parchment in an obscure library, Rome would still have burnt. People would still have died.

And if the scientist had gone to the Coliseum[3] while a man was fighting a lion with his bare hands and the crowd was transfixed (roaring and cowering and cheering) and this scientist had stood to the side saying studiously, "Nero is planning to destroy the city of Rome by fire", the result would still have been the same.

***

Let's return to Nero and his violin. The violin he never could have played because it had not yet been invented and would not be invented for another fifteen centuries. As we will see in the chapters to follow science is able to place the question of Nero and his violin at the very left of the axis of truth. It is 'definitely untrue'. To these scientists, that settles the issue. The falsehood has been discovered. Determined.

But was Nero present during the fire? This question proves to be less clear-cut. And the evidence to determine it conclusively will probably never be found. The question is doomed to float forever between the poles of 'false' and 'true'. It is 'possibly true' and 'possibly false'. This means that for the time being it is 'more true' or 'less untrue' than the assertion that Nero played a violin while Rome burnt. In a world where truth and falsehood is determined on one line, it would look like this:

But we all know we don't just believe things in this way. There is a second line that runs in a different direction. The line of 'beauty' and 'ugliness'. Or we could call it 'captivation' and 'the commonplace'.

We go to a magic show and the magician draws a rabbit out of a hat. Turns a playing card into a dove. We know it is untrue. Do we leave? Do we stampede for the exit? No. We stand in a long queue to get in. We part with salary or pocket money or pension to get a ticket. To see him capture our imagination. We clap and shout for more.

Or we go to a football game, to see a winger running at full speed, kicking a moving ball a hundred degrees to the left of where he is going, at exactly the right height, with spin and thus swerve, exactly between the defender and the goalkeeper. For a diving header by the onrushing striker, into the corner of the net.

We go to the circus to watch the trapeze act. The acrobat letting go at exactly the right moment and the correct angle. Making a double summersault and being caught by the catcher on another swing, moving in another direction at another angle.

The more unusual, the more skilled, the more impossible, the more dangerous, the more ground-breaking, the more appalling – the more it transfixes us. The more 'beautiful' it is. We love it and internalize it completely, irrespective of whether it is true or not.

Now, on this line, let's place two things, words and an image: Nero playing the violin while Rome burns and the statement "Nero was absent while Rome burnt".

There is no doubt that the notion of a violin against the backdrop of a city ablaze and the citizenry in panic is more captivating than the question of whether Nero was indeed present or was perhaps standing in Antium with a sharp stick, trying to pick at morsel of food stuck between his teeth. The violin is more captivating.

***

Truth, like lies, can be beautiful when it is told in a way that is captivating rather than commonplace.

This is demonstrated by the lie of Nero and his violin. If we combine the graph of 'Lies/truth' with the one of 'Beauty/ugliness', it looks like this.

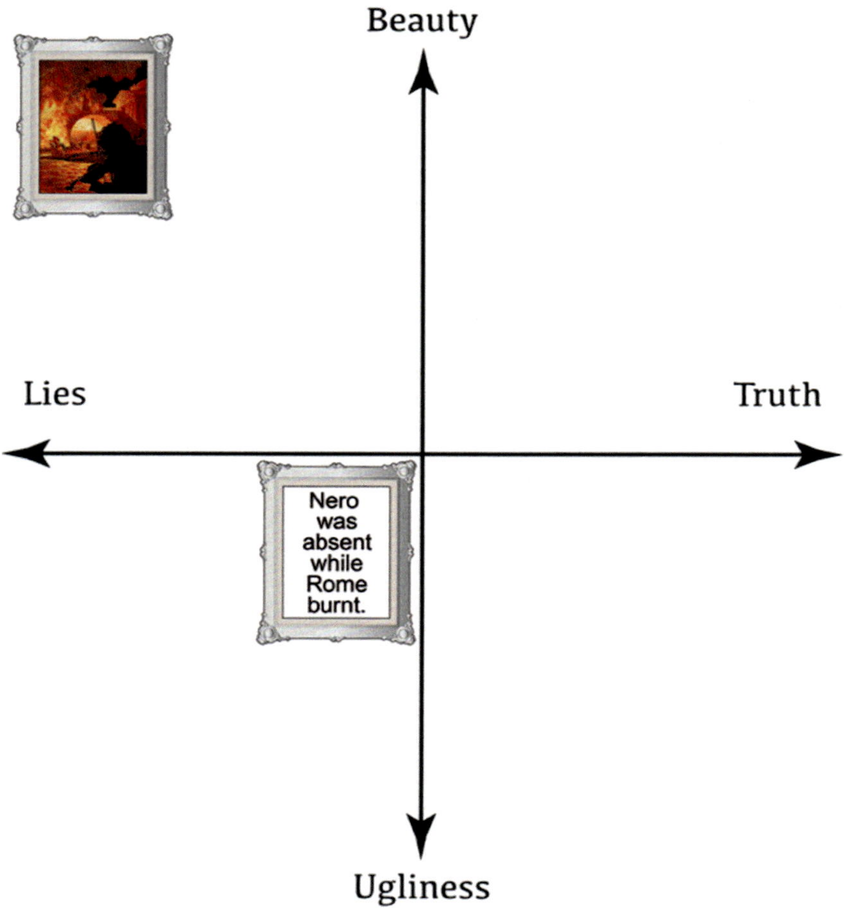

On this graph, either the question of the violin or the question of Nero's absence from Rome during the fire can hold a 'higher'/'truer' place in our minds, depending on the way our brain works. If we are very logical, like a computer that knows only zeros and ones, or if we work in history, or work in music, we operate on the horizontal axis. We remember the violin story as 'untrue' and the question of Nero's absence from Rome as 'more possibly true'.

More true.

But if we're more interested in captivation than truth, more interested in happiness than sadness, we operate on the vertical axis. We remember the violin and Nero playing it. It sits 'higher' on the graph. We remember the insanity in his eyes, the panicked crowd, the unstable falsetto voice. We pay no attention to the commonplace question of whether he was in Rome at all.

The perplexed scientist on the yacht, weighed down by the report of an approaching cyclone,,doesn't know this, but our truths need to be beautiful – or they cannot compete with beautiful lies.

***

There is the story of the yacht. Six of us in the middle of the ocean. Needing to change course, but not in agreement. That story will unfold across the book, and its end will be the end of the book. How do we perceive and understand? How do we communicate? Can we convince others to act in a specific way with us, so that we survive? What happens in the end?

Then there is the story of our blue planet, the yacht in the sea of space. Millions of species, billions and billions of organisms. A coming storm that science, standing at the radio, knows about. An equally pressing need to change course – but we're not in agreement.

How do people perceive and understand? How does science communicate? How can science convince people to act together in specific ways, so that we can survive? That story too, will be told. How science has stood on the deck proclaiming an approaching storm over and over again, with little success. Yet repeated the identical strategy – like Sysiphus rolling a boulder up a hill, for all eternity.

We'll inquire into the nature of the approaching situation that we are likening to a storm, and say: "What is this imposter?" We'll find out why science is the way it is, and communicates the way it does. How people really listen, and how little science allows for this.

Noticing how a lack of agreement is holding us back, we'll map the DNA of our world: Is it all competition? Survival of the fittest? Red in tooth and claw? Or all harmonious, the way John Lennon and his generation imagined it?

And of course, we'll investigate what the people on our blue spacecraft do in the face of our inertia and our disagreement about what route to follow. Some try to convince in the old way again. Some, aware that they can't leave, protest peacefully. Some are ready to use violence to bring reason to our actions. As we vacillate, the storm keeps rolling in.

There is the story of truth and lies, divorced from beauty and ugliness. The story of Socrates, the man who set science on the road to becoming what it is, and communicating the way it does. One dimensionally. Socrates who set up the scientist on the yacht to proclaim the ugly truth about the approaching storm to the others on the deck in the vain belief that they will process the information with perfect rationality, like a computer. All zeros and ones. Socrates the thought leader of his people, the intellectual giant who was supposed to keep them on the right path but preferred to let them stray – so history could show him to have been righteous, and right. Socrates, who'd rather die than acknowledge a multi-dimensional world of conviction. Just like science today.

The fourth thread is the tale of Nero, the emperor who cared nothing for truth and lies, but lived only for art and captivation. Like Socrates, his world was one-dimensional – but he chose the other axis. The vertical one of beauty and ugliness. And because he cared not for truth and lies, he didn't care for right and wrong either. Whatever was captivating was right, whatever was drab was wrong. This failure cost him everything.

Socrates knew only truth, Nero only beauty, but to survive on the yacht and as a planet, collaboration and competition must find a balance. People must come together, science and art must be rejoined, the ultimate norms of Socrates and Nero have to merge.

To give us beautiful truth, and conviction.

PART 1

Beautiful Lies

1

Believers and non-believers

A minute after Sam dismissed our warning, the two of us were back at the radio. Without having agreed to meet or go inside. The cyclone warning had come and divided the boat into believers and non-believers, adherents and agnostics. Instantly. And somehow recognizing this, the two of us had gone back in. Adherents huddled together next to a VHF radio with limited reception that somehow unites us now.

"What just happened?"

"I… I don't know."

"But you believe me?"

"Sure I believe you. I wish I didn't."

Outside, it is still just the swell and the humidity, overwhelmed by the sunshine and the absence of wind. Just a perfect, calm day. Mocking us every moment with a loud, "Told you so!"

"He's your husband, Julia..."

"You'd think that would count for something. Make him believe me. But he's outside, lying in the sun, and I'm here, and it feels strange."

Our voices are low, as if we're afraid some enemy will hear us. An enemy including a spouse. "We should have talked some more to all of them, before Sam lay down again."

"You didn't say anything, Juan-Franco. Nobody even knows you agree with me. They probably think you're trying to restore reception ..."

That suddenly seems like a good thing. We don't want to look like conspirators. Don't want the "us" and "them" to settle into fixed formations that cannot change.

"I'll go out," Juan-Franco agrees. "Whatever happens, remember that I'm behind you. But for now I won't say anything."

He stops and turns around again, his skewed upper body etched against the bright light outside. The muscled left side, the atrophied right. "What's the barometer doing?"

"It won't be doing anything now. It'll stay flat and then fall like a stone in a day or so. Feel the humidity? See the swell? That's not small. It's coming. We can't wait for the barometer. By the time it drops we'll be too far gone. We have to get away from the reefs. Away from the island. If the storm hits us it will drag us any which way. We might survive that if we're not near the eye and if we're very lucky. But not if we're near the reefs."

***

Juan-Franco walks out. His legs are the same, both strong and tanned. He's wearing shorts and a long sleeve cotton shirt. The others have only ever seen him in long sleeves. They know about his accident, when he dived off another yacht after Esthia. Into shallow water with breaking surf and coral reefs. People talk.

Angela watches him emerge. She's enjoying the sunshine, but the swell is becoming uncomfortable and she wonders if she'll be queasy soon. Or is she worried about what Julia said about the storm? She looked so concerned as she spoke. And Sam so sure that Julia was making a mistake. This is why Angela watches Juan-Franco. He's been inside. Is there radio reception now? His face doesn't speak of success. But is he worried? Does he believe Julia? Angela can't tell. The sun is so bright she has to squint. And it's so humid she has to wipe her brow to prevent sweat from running into her eyes.

For a moment she imagines Julia to be right. The huge storm to be bearing down on them. And how frivolous they must look, lying in the sun, working on their looks, thinking about that first Mochito in an hour or so.

And likewise, how silly to toil away preparing for a storm report that was meant for Malaysia but received here in error! Angela scans the horizon and sees not a cloud anywhere. Just nothing.

It's a perfect day. No doubting that. So take it easy. She sees Sam flat on his back, sunglasses making it impossible to confirm that his eyes are closed. Perhaps he's sleeping. Relaxed. Looking forward to arriving on the island. Angela covers her head with a blouse and lies back.

Lorenzo puts his hand on her knee. Angela exhales and starts to enjoy the sun. Lorenzo is there. They're always close to each other. He looks out over the swell and remembers the time a school friend with a car raced home drunk and skipped all the red lights. How helpless he felt in the passenger seat, how completely delivered to fate. They'd hit something or not. Be injured or not. Die or not. But someone else was driving and it was too late to get out of the car. Was it the same? Maybe. Impossible to know. A storm might be coming and it might be nasty. Or not. Why prepare for something that will never come? Just enjoy the day and take steps when it's clearer what's happening. If clouds come or the wind picks up, they can decide what to do.

Erin is reading a novel by a female writer she's never heard of. She forgot to bring a book and grabbed one at the airport. The writing is fast-paced, irreverent and funny, set in New York. An American Bridget Jones, Erin thinks. She knows the plot will follow the Bridget Jones series, but she doesn't care. The predictability gives a certain joy. She gets up and goes inside to make herself a Mochito.

Julia is at the radio, fiddling around. Whipping the dead horse. They exchange pleasantries and Erin goes back outside, Mochito in hand.

2

Science's most beautiful lie – "Truth speaks for itself"

When Julia was young, she lived in a small bedroom on the seventh floor of an apartment building. An unfashionable one in stark grey, its concrete exterior unpainted, probably built in the early seventies. It was in a neighbourhood of stark and practical buildings inhabited by stark and practical people, or people who lacked the means to be otherwise.

Julia's father was in medical research. Her mother, before she was born, had been a musician. Had busked in subways and been to San Francisco that summer when everyone was there; she would have been at Woodstock if a broken radiator had not intervened. Julia often remembered her with flowers in her hair. And in her own. Her mother had always put flowers in her hair when she was small and told her that some years before, almost everyone wore it like that. That it was a good custom to observe and pass along.

"There is a lot we can learn from flowers," she said to Julia one day, stroking her forehead. Her voice was wistful and her eyes looked at something far away. It was in the city park. Late summer, not warm or cold, but green, and there were people on bicycles and kids on tricycles and some dogs on leashes and behind them the people they were taking for a walk. "Flowers are here today, gone tomorrow...", her mother continued. "They don't leave footprints. They are beautiful and bring joy. And they disappear without complaints."[4]

But her father didn't think flowers were so important. He got angry if he arrived home from work and found that Julia hadn't been bathed yet, or fed her supper. He thought there was a time for these things to happen and raised his voice and opened his eyes and his mouth wide as he spoke. More so when Julia's mother had injected herself with the medicine she took sometimes and happened to be drowsy or sleepy. Once or twice on these occasions, he shook her, and her head rolled on her shoulders like that of a rag doll.

"It's to make me more like a flower," she said to Julia sometimes as she tied a string around her arm before the injection. "It is medicine for pain. It makes me dream and forget and not complain."

But then she'd be very cold when she woke up and sick sometimes. Quiet and sad. And her father would speak sternly with her mother, in the same voice he used to speak to her when she was naughty, and explain in words Julia didn't understand, and throw away all the injections and all the medicine.

And then that day when it was raining so much outside her mother had slept and slept with the needle and syringe still in her arm until the doctors came and put a blanket over her face and took her away in the back of a white car.

After that Julia grew up in the same practical and stark apartment block, in the same bedroom, but only with her father. From her bedroom window, if she looked down and to the left, she could see the very edge of the city park sticking out behind another grey and stark building. She could sometimes see a bicycle and rider leaving the park. Sometimes a dog, pulling someone out of the park and back towards home. She positioned her desk where she had the best of this view and often stopped concentrating on her homework and looked down and thought about flowers.

A grey year passed, including a winter when the days were short and few people went to the park. When dogs and owners walked hunched, as if to just get it over and get back inside. In the spring on a late afternoon, as she looked away from her homework again and towards the park, she saw movement across the bottom of the window, too quick to detail. She got up and looked over the sill, and smiled as she saw a swallow's nest. It had been built against the sheer granite, defying gravity and the smooth surface. A swallow had to be inside – or perhaps more than one – large and small.

She kept looking out for the swallow after that, using her peripheral vision to monitor movement while she did her homework. Rising occasionally to peep, even as she knew nothing would be visible. But the dark flash would sometimes come from somewhere and disappear inside. Or emerge and fly off. Giving her a little more time to watch and try to see what her swallow looked like.

After that, it was as if the park had come around the corner. Up from the street, past the second and fourth and sixth floors, all the way to the bottom of her window sill. As if the building was no longer grey and stark, because now it supported a swallow's nest.

Her father still reminded her of the importance of detail. Procedures, time schedules, routine, discipline. She listened, worked hard, excelled at school. As the days shortened and the fallen leaves began to block the gutters, she knew the swallow would be leaving soon. There wasn't a way to say goodbye, wasn't a way to know that it was setting off for Africa at this hour of this day – and watch it go. The going would be manifested only in the absence of movement in and out of the nest. And when for two days there was nothing, she understood that the swallow wasn't moving just when she happened to be looking in another direction, or to be away from her desk. The swallow was gone on a journey of many thousands of miles, and might come back, or not.

It came back in the spring, and the next spring, and the next. Six years in succession. Whatever else might be happening or not, the swallow was there.

Then, the curtain rail in Julia's bedroom had to be changed, and her father came to drill holes in the wall. It took a long time, and there was a lot of noise and vibration. For the next few days, she didn't see the swallow. But this had happened often over the years. She'd sometimes see the swallow as much as three times in one day, then not for several days. But then it became too long, and Julia became irritable without knowing why. The next morning when she looked over the sill, the nest was gone. A few remnants showed where it had attached to the granite for six years or more – but that was it. Like a flower, it had come and gone, and left almost no footprint.

Hands on her shoulders, arms crossed, head leaning against the window, Julia closed her eyes and imagined how, years ago, the nest had been glued to the building so tightly that not even a hurricane would dislodge it. And how, through hot summers and blustery winters, a small crack must have emerged, away from sight and mind. Grown by a mere millimetre at a time, every few months, with no impact whatsoever on those in the nest. How, with gusts and movement inside the nest the crack had grown until the nest was hanging by a thread. And how her father had drilled holes in the wall and sent tiny vibrations past the nest, straining the thread until it broke and sent the nest tumbling down to the street below.

And Julia remembered her mother and the story of the flowers and also her father and the importance of discipline and routine. Should she have put a broomstick out the window every winter and gently prodded the nest to see if it was still secure? If she'd done this, it would have fallen the last winter, and the swallow would have come back and built a new strong nest, in the same place or in another. And would have survived ... not maybe, but definitely.

And so, seeing out her teenage years without the company of the swallow, Julia became a scientist wanting to study birds' nests that may be about to collapse. And similar things. And just as her father had wanted her to, she began to look at things in great detail. She stressed procedures and routine. Time schedules and discipline.

A bird's nest can deteriorate for many years without any visible effect. And then a tiny vibration can send it plummeting. But careful observation and occasional testing could have prevented it.

Julia became a careful observer. She also learnt to drive. Where she grew up and spent most of her early life, far from San Francisco and Woodstock, people drive on the left hand side of the road. Therefore, keeping left is a very important thing to master when learning to drive a car. It's what prevents horrific, head-on accidents. Always drive on the left. When approaching another vehicle or a pedestrian or an elegant horse rider in a narrow country lane without any marks indicating the middle of the road, always veer left. Even when she was small, when her mother was still alive and she sat alone in the back seat of the car with her parents in the front, she would see thousands and thousands of times how her father always kept left. A cautious driver on the highways he'd keep left so the faster cars could pass on the right. And going up escalators in the city, Julia learnt to stand on the left if she was at ease and to walk briskly up the right if she was in a hurry. If another person pushing a trolley approached from the front in a supermarket, both Julia and the other would veer left, and avoid a collision. Keeping left became a default setting, something she and all the other people in the country of her birth did without thinking. Even if a crazed person in a park should throw a knife at any of them, they would almost certainly take evasive action by diving to the left.

Why left? It must have come from somewhere. Well, in feudal times, mostly right-handed knights carried their swords on the left, so they mounted their horses from the left, pulling over to the left to dismount so they were off the road. Knights on moving horses, approaching from the front, passed stationary hoses and knights on their left. And logically, horsemen passed those moving in the opposite direction on the left. In addition, the majority carried their weapons in their right hands and when mounted and charging at an enemy, would thus veer left to attack more effectively. More than a thousand years earlier, probably for similar reasons, the Romans drove on the left. And so, what the Romans did two thousand years ago would ultimately cause Julia to default to keeping left, wherever she went, without thinking.

And similarly, Julia communicates in a specific way about facts. When walking out on the deck of the yacht, concerned by what she's just seen, her body language might speak of anxiety through the brisk walk, approaching a jog. Her words, however, because she allowed them to come out without planning, were the words and the tone of a scientist. Measured, stripped of adjectives, factual.

This too, has become ingrained in her over many years in her chosen profession. The conventions she follows likewise go back very far, beyond the Romans to the conditions shaping early mankind. They were espoused and reinforced by a stubborn Greek named Socrates, who loses his life for wanting truth to speak for itself. Despite this, the conventions of scientific communication are amplified by the persecution of scientists in the Middle Ages and come to full fruition and exposition in about 1800, with the Enlightenment. The conventions are then perfectly applied by a visionary scientist called Ignaz Semmelweiz fifty years later, a man who also loses his life for his diligence – much like the early exponent, Socrates.

And even after that, today, scientists like Julia continue to follow the same conventions. Strip it down to the facts. Make it cold and hard.

On the yacht, these conventions now threaten the lives of everyone.

***

We all speak about progress from time to time. Human progress. Some refer to 'civilization', others talk about 'evolution'. Most often, people talking about this progress refer to a moving away from the natural state of things, the baseline of where it all began. If they look at the natural world, they see a pride of lions pulling down an antelope and devouring it while it's still alive. The heart still beating, the wide open terrified eyes, the blood everywhere, the sounds of feast and dying and last-gasp pain all mixed together. It seems incredibly wild, cruel and callous. People see the relentless competition for survival as the DNA of a natural world that is "red in tooth and claw".[5] In human affairs, this primeval state equates to "might is right": The caveman with the biggest biceps is King. The giggling and frivolous queen requests the head of the prophet on a silver tray – and gets it. The father beats the child for arguing with him, even when the child is correct. Don't talk back. Know your place. "Didn't I warn you?"; "I'm not saying this again!"

The humiliated and sore child lies in a dark corner, sobbing not just about the rejection and pain but about the utter injustice of it all. He was right but his father was bigger and stronger and held the whip. And wouldn't stop thrashing him until he stopped arguing back and cowered in total obeisance before the master of this little corner of the universe.

It's all about power. In that world, women are the weaker sex and should know their place. Children should be seen and not heard. Animals are lower order beings and can be utilized, eaten or abused, as the whim of the strongest man may dictate.

What can the child do? Eat, exercise, grow strong, learn to be fleet of foot and fist. Wait for the father to weaken. For the self to grow stronger. For the right moment. Then thrash him. Annihilate him. If necessary, kill him. Become the master of this universe, crush opposition mercilessly, survive for as long as possible. Then die by the sword and the whip, as he's lived by the sword and the whip.

When thinking about progress, people often mean divorcing humankind from the DNA of the world. They see humankind as having, by necessity, started there – but also as having progressed through evolution towards an ultimate end goal, namely complete harmony, collaboration, equality, acceptance of diversity, peace. John Lennon's *Imagine* is perhaps the best manifesto of this dream:

> ...imagine there's no countries
> It isn't hard to do
> Nothing to kill or die for
> And no religion too
> Imagine all the people
> Living life in peace
>
> You may say I am a dreamer
> But I'm not the only one
> I hope someday you'll join us
> And the world will live as one
>
> Imagine no possessions
> I wonder if you can

No need for greed or hunger
A brotherhood of man
Imagine all the people
Sharing all the word...

You may say I am a dreamer
But I'm not the only one
I hope someday you'll join us
And the world will live as one[6]

In this world, the concept of "might is right" ceases completely. Physical strength or sophistication of weaponry no longer decide who is right and wrong. Rationality and good sense regulate the behaviour of everyone, all the time.

The thrashed child, sobbing in the dark corner imagines a universe of 'right is right'. A place where there is a way to determine whether he is correct or incorrect in the statement that he is making to the father. And where truth is king. A place where the father patiently listens to what the son is saying, asks questions, challenges gently. The son gets the feeling that he is only half right and goes away to think, then returns with more nuance and confidence. And ultimately the father smiles. Pats him on the shoulder and says: "You are right."

Idealistic people see humankind as being on a journey from one mode of being to another. From 'might is right' to 'right is right'. When they hear someone saying that women are the weaker the sex, they call the utterance 'backward', because they see it as coming from someone being left behind on the journey, someone closer to the beginning than the average human. They might call someone displaying racism a 'dinosaur',[7] implicitly because s/he should have been chosen for extinction by evolution already, for their inability to adapt.

This is the context in which the conventions of science evolved – the conventions that cause Esthia to be pleased that she doesn't have to sell science. The habits that caused Julia to communicate the news of the approaching storm in a very factual way. All scientists were children once, most ran across an authoritarian somewhere in childhood, many were thrashed for speaking their minds even when they were right. From the earliest times of human history, clever but weaker people must have been wishing for a world where people would speak rationally and logically, where their words would be heard rationally and logically and objectively, where in each instance perfect input would lead to perfect output, perfect reasoning to perfect conclusions, perfect presentation to perfect persuasion.

***

It's mid-afternoon in the early summer of the year 399 BC in Athens, and the trial of Socrates is winding down. The people of Athens have ostensibly charged him with corrupting the youth and denying the existence of the gods, but actually it was just a few conspirators conjuring the best charge they could think of.

It would all be settled by sunset. Socrates has already been convicted and it's only about his punishment now. He knows many of them want him dead. They observe him, waiting. Some look outside, a few yawn, one or two seem to listen. They all know the rules allow him this moment of attention. He can make his case, but the privilege would pass and then their turn would finally come. He's spoken for a long, long time. First in his defence, now on what an appropriate sentence should be.[8]

Still he continues to bait them, provoke them, insult them in the most discreet and subtle of ways. Still he leads them steadily towards killing him so he can prove their base natures and their lack of virtue. So that in death, he can win for all eternity. He's not talking to them, he is talking to posterity, carefully making his case, and leading his detractors to every possible blunder in making theirs.

"The world has found that I am in some way superior to other men...," he says along the way, without a hint of bashfulness. And then he reminds them of all their evil deeds, and how he tried to stop them, and how he had nothing to do with it. He sees the narrowing of eyes, the clenching of jaws.

"Yes, that's the way you are," he thinks as he speaks. "And always have been. Utterly variable, utterly lost without me."

Just before they convicted him, he referred to the many men who've stood in the place where he now stands, accused and possibly facing death.

"They bring their children and cry many tears and talk about how their wives need them," he said with scorn. "Hoping to be spared – as if they are immortal. And I too have a wife and three sons. But speaking like that would discredit me and you and indeed the whole country. There's something wrong in putting up a show instead of just informing and convincing the judge and jury."

He said then he would only give them cold reason, and they reacted by convicting him, for they wanted him dead. Socrates, their conscience.

Now again, he reminds them: "Remember that I don't stand convicted because I didn't tell you the truth or give you a full account. No, I am not convicted for too few words. I'm convicted because I didn't talk to you the way you wished to be spoken to. I didn't weep and wail and lament and put up a show. And let me tell you clearly, I'd rather die speaking only the truth than bring you the circus and live."

He's put his finger on it. He's been insisting that truth speaks for itself. That was what it was all about, ultimately, and he was happy to swim with the cold truth, or sink with it.

He would die soon afterwards, but he would die believing that he'd shown up the folly of the mob. That in death, he's won. And his words and thoughts would spread far and wide, far beyond even the world that was known in his time. His naked truth would become a mantra, and millennia later would affect a woman named Julia on a yacht. Cause her to relay a dire warning in a certain way.

And indeed, the stubborn Greek influenced many other clever people living in that same era, and caused them to communicate other, dire warnings while refusing to include anything but the cold, hard facts.

If on that day, Socrates could see the possible consequences, he may have spoken otherwise, and may have remembered that the mob needed him to guide them – not to lead them astray to win a point.

Then too, he may have seen that the route he chose didn't lead to victory.

But Socrates would die believing that in death, he won. That he has made the case for cold and hard truth against the inanity of the mob. That the mere fact that they chose to execute him would show that he was right, and they were wrong.

For this, he's been lionized. The man who was willing to die – if that was what doing the right thing required. But did he change the behaviour of Athens and her people? No, he removed his own voice of reason and wisdom from the equation and allowed the scales to balance the other way. Socrates tacitly accepted the probable ruin of his people and morally jumped ship at the last moment, accelerating ruin but making sure it didn't occur in his name. When he saw how the stories and tears of other convicted persons swayed the jury, he looked at it in contempt and said: "This is not the way humankind should be. We should be perfectly rational. Given the facts, we should be able to come to the right conclusion. I will not adapt to how humankind is but will insist that humankind adapts to me, the perfectly rational one."

For Socrates, it was human nature that was the accused in his trail, accused of illogic and utter variability. He was the judge, weighing it. And he duly convicted it and likewise sentenced it to death by determining it to be an utter basket case, terminally flawed.

Since then, people have had two options. The first was to conclude that Socrates was right about human nature, that people are as often emotional as rational, that laughter and tears weigh as much as logic (or more), that stories weigh as much as rhetoric (or more). Communicate with people as they're actually wired and have a chance to convince them.

The other option was to conclude that Socrates was right about human nature but to believe that Socrates constituted an early example of a human that has risen up the ladder of evolution to perfect rationality and that our task (and indeed the task of progress and civilization) is to pull all of humankind up along that ladder until we're all exactly like Socrates. Every man, every woman, every child. The busy baker, the stressed hairdresser, the lawyer with the fragile self-esteem, the simple stable boy, the megalomaniac president. The vain beauty queen, the paranoid schizophrenic, the victim of child abuse. The coke addict, the alcoholic, the religious zealot, the double amputee with the broken mind. To believe that, if all of us are not always perfectly rational, it is just a function of time. There hasn't been enough time for all of us to become Mr Spock. Evolution hasn't had sufficient opportunity to deliver us there. Give us time, we'll arrive. And we have to start somewhere: so let's imagine we're already there. Let's not pander to the backwardness by tipping our hat to emotions or by telling stories. Socrates has already poured scorn on such people. Let's just communicate with everyone as if they're Socrates. Better to communicate with perfect rationality and fail, than to acknowledge human nature for what it is and succeed. For that would consign human nature to invariability – and be a bigger defeat.

***

History after Socrates continued to be human and messy. Athens started a steady slide into obscurity. The Greeks lost their pre-eminence in world affairs and gave way to the Romans. After Nero the Romans, utterly devoid of any moral compass, themselves continued apace towards a self-created abyss. As they imploded and fractured, the Middle Ages commenced and the clergy became the father holding the whip. "I know best. Don't talk back. Know your place. Didn't I warn you? I'm not saying this again! The Inquisition may have to look at this".

There were witches and witch-hunts. When uncertain whether someone was a witch, she was thrown into the water with stones around her feet. If she was a witch, she would defy gravity, not drown and come to the surface – and there would be proof she needed to be burnt.

Witnesses in legal proceedings were tested for veracity by soothsayers, who would knowingly move around the witness with much ado to pick up a 'sense' and ultimately declare through their divination that the testimony was truthful or false. Accused persons were subjected to trial by ordeal. We think you're an adulterer or planning treason. Pluck this stone from a pot of boiling water and if you're innocent, God will protect you from injury.

Anyone could become victim of this system. Initially, there was no specific reason to pick on scientists – perhaps because scientists were in short supply for those ten centuries. Other career choices were more popular. But the wheel had turned well away from where Socrates had wanted it to go. "Might is right" was ruling the day and for those at the top of the pyramid not actually believing in soothsayers and trial by ordeal, rationality was cynically being abused to destroy opponents and stay in power.

In the minds of leaders of the Church, it worked like this: there was a God who created the universe and had appointed them as his direct functionaries. Only God was above them. Below them was all of humankind (who could only approach God through them), then the animals. Naturally, in this giddy scheme of things, Earth was at the very centre of the universe and everything revolved around it. At the centre of Earth was mankind, God's special species created specifically to rule, and at the centre and the top of mankind, the clergy. Ultimately, this implied that the entire universe revolved around the upper echelons of the clergy. Any notion that the centre of the universe might be somewhere else would threaten the entire set of assumptions that the power relations were built on.

Nearly two thousand years had passed since Socrates and his Trial of Human Nature. Human rationality was on the retreat for most of that time. Even amongst the Greeks, the idea that the sun was the fixed point around which things revolved was known, but now, under the Church, this was unthinkable. If evolution was steadily working towards the epoch of reason, it had to be on a timescale of millions of years.

But science was stirring. People were becoming inquisitive and asking questions again about the world we live in – how it works. Travel was increasing, ideas crossed boundaries and were cross-fertilizing internationally. By 1514, Nicolaus Copernicus, a Polish scientist, was sure that the earth revolved around the sun. But he was also involved with the Church and wise enough to delay the publication of his work repeatedly. Only just before his death in 1543 did he allow publication to proceed. It's rumoured he only saw the printed copy of his life's work briefly, on his deathbed. He'd dared to challenge the basis of just about everything, but died before he could be criticized or persecuted.

In the century that followed, the brilliant Italian, Galileo Galilei, was less wise. He had an early run-in with the Church when it appeared the theory of Copernicus (as he espoused it) might cast doubt on biblical verses suggesting the earth to be at the centre of the universe. He got off lightly and was ordered not to advocate the truth of the so-called heliocentric theory. After some years of successfully avoiding conflict, his convictions overcame his better judgement and he published a book on the topic that featured a simplistic person saying the very things the Pope of the time was saying about the matter (the earth is at the centre of everything, and other, related beliefs). In Galileo's book, this simpleton of his creation was made to look silly by the truth. Predictably, a full investigation by the Inquisition followed and Galileo narrowly avoided torture. He ended up recanting his opinions and spending the remainder of his life in house arrest – with his writing banned.

And yet, just as the grip of both Athens and Rome on power eventually slipped, so too the power of the Church was slipping. By the time the Industrial Revolution was in full swing a hundred and thirty years after Galileo, everything was changing. Science was behind the wonders of the new machines suddenly doing so much, so effortlessly. There was urbanization, a growing middle class, social disruption. Science had escaped from its bonds, and scientists had stopped fearing reprisals for mentioning the truth. Within another hundred years or so Nietzsche would be able to publish the catch-phrase "God is dead" – and fear no retribution. [9] Cross-examination would replace soothsayers in legal proceedings and accused persons would be judged on the weight of evidence, not by ordeal.[10]

The might of the Church had dissipated and a window of opportunity had opened up to try again for 'right is right'. Appalled by the centuries of backwardness and repression under the Church, science said: "We're winning. This is our time. But we must never, ever be like them."

In the time when the universe rotated around Earth's clergy, if Julia appeared in the deck of a boat and said to the other five people: "There's a storm coming!", what would have happened? If she commenced by saying: "Last night I had a dream and the Mother of Jesus appeared to me with tears in her eyes and a compass in her hands, telling me that all of us must avoid the fate of Jonah", then there would have been serious contemplation of whether they were running from their destiny and God's command and whether they were likely to be caught in a storm as a result. They may have ended up changing course. But if she claimed to have heard it through a machine that allows people to hear others speaking from beyond the horizon (a radio), informed by someone who had other machines that could look into the future to predict the weather and the seas (a weather forecaster), she was likely to be disbelieved, suspected as a witch and told that these things are all in God's hands anyway.

Science wanted to remove subjectivity and arbitrariness from the decision-making process, because for thousand years a person in power could just say, on the strength of little or no evidence: "I think you're a devil-worshipper" and the person would face death unless they were willing to pluck a stone from a pot of boiling water without showing pain. It was a win-win outcome for the one holding the whip and a lose-lose for anyone who dared to anger him.

And so science created the world of Socrates, where truth and lies are separated on a single line, in one dimension. It created a set of basic rules saying that the truth has to be told like Socrates told it; just the facts, no adjectives, no emotion, no juice.

Never mind that it once cost Socrates his life and sped up the decline of his civilization. Just give us the facts; cold and hard.

Their cold contrast with the warmth that might be present around a camp fire, or friendship, or pride at a relative who has excelled. And their hardness is 'hard' in the way prejudice is hard, 'might is right' is hard, the rejection of another person is hard. Not soft like forgiveness, or diversity, or love.

***

It was going to be the Age of Reason. 'Right is right'. Enlightenment. The time when weaker but cleverer people would finally prevail in discourse between the weak and the strong.

There was so much excitement – the possibilities seemed endless. In France, the idea of an encyclopaedia containing all important knowledge, took root. Over about two decades a multitude of contributors wrote material for the *Encyclopédie*, which sold about 25 000 copies inside and outside France and must have made an enormous contribution to the dissemination of knowledge. The main contributors included Voltaire, André Le Breton, Denis Diderot, Montesquieu, Jean-Jacques Rousseau and Chevalier Louis de Jaucourt.

In the United Kingdom, Adam Smith was laying the foundation for free market capitalism; while Francis Hutcheson promoted the Enlightenment in general and especially liberty from political oppression. James Watt invented the steam engine.

In Germany, Christian Wolff and Immanuel Kant were actively investigating the Enlightenment from a philosophical perspective. Frederick the Great, the ruler of Prussia, became a patron of the Enlightenment and, amongst many other initiatives, invited Voltaire to reside at his palace.

Reforms were introduced in Russia and Spain under Catherine the Great and Charles III, and also in Poland.

In America, prominent supporters of the Enlightenment included Benjamin Franklin and Thomas Jefferson. Enlightenment ideas were manifest in the American War of Independence and Bill of Rights and ultimately spawned constitutional democracies in Canada, Australia, Botswana and South Africa (amongst others).

As machines were invented to do what had never been conceivable before, the future seemed limited only by imagination itself.

And yet – it was ugly. Unaesthetic. From the simple and largely oppressed life in the agreeable and picturesque countryside, people had migrated to sprawling new cities where most lived in squalid conditions without a blade of green in sight. There was child labour and prostitution and much else besides. Fossil fuels had been discovered like "a huge treasure chest in the basement"[11] and were being combusted on large scale to drive the machines doing the production. There was much more pollution.

Soon, there was oppression; not by the church or by a monarch, but by the mob. In the 1790s, tens of thousands of Frenchmen and women lost their heads to the guillotine, during what was called the Reign of Terror. 'Pure rationality' could still lead to horrific regimes – by the twentieth century it would do so on an unimaginable scale.

Reason increasingly meant the absence of emotion not only in a moment of decision-making about something scientific – but about life in general. Poetry became very intellectual, looking for objective truth, steering away from emotion. The following is a fragment from Alexander Pope's *Essay on Criticism*:

...Name a new play, and he's the poet's friend,
Nay show'd his faults – but when would poets mend?
No place so sacred from such fops is barr'd,
Nor is Paul's church more safe than Paul's churchyard:
Nay, fly to altars; there they'll talk you dead;
For fools rush in where angels fear to tread.
Distrustful sense with modest caution speaks;
It still looks home, and short excursions makes;
But rattling nonsense in full volleys breaks;
And never shock'd, and never turn'd aside,
Bursts out, resistless, with a thund'ring tide.[12]

Yes it's clever and yes – he has command of language. But why would we read it? To marvel at the intelligence? How would it inspire us?

By about 1795, the way Julia would walk out on the deck and speak about the approaching storm had been hard-coded into the scientists of the future. The scientific method was in the ascendancy – launched by Descartes and Newton and entrenched in the late eighteenth century by the scientific efforts at empiricism.

It asks the son, before he speaks, to eliminate all personal bias, all emotional involvement, all prior loyalty. It then asks him, if possible, to measure what he is talking about. Or to find a way to test it. A way that would show it up if it were wrong. And then to tell everyone how he's discovered what he knows, so someone else can repeat the same experiment and confirm the same answer. And the son is asked, finally, to make predictions that depend on the truth he has discovered. If the predictions come true, he knows it to be true, and all the world of "right is right" agrees with him.

Science has found a way, when it concerns things that can be measured, to ensure that most of the time, speakers speak truths.

Science is quite good at policing the speakers, but totally inept at policing the listeners. It has assumed that because truth is spoken, truth will be heard. This is the lie. For not all who listen believe that "right is right". And very few who listen leave behind their personal bias, all emotional involvement, all prior loyalty – even sometimes when they are from the world of "right is right". Even sometimes when they are scientists. The beauty of the lie is in the dream that this success of recognizing truth in speakers could be expanded to all speakers and all listeners, all things whether measurable or not, and on all occasions.

There was a notable and immediate reaction from those who believed in the importance of beauty and captivation; they changed the poetry and literature of the time but had no effect on the scientists because the arts and sciences were drifting inexorably apart. The marriage of science and the arts that had existed for millennia as dual ways to understand our world had finally hit the rocks, and the differences had become seemingly irreconcilable. In the ensuing separation, science claimed the exclusive ability to plot truth and falsehood on a single, horizontal line. The arts lost their interest in objective truth and walked away with the ability to create beauty and to captivate.

It wasn't immediately obvious, but the one had retained the means to discover objective truth; the other the means to convey it. Sell it.

And consequently, while the separation persists, certain truths have become virtually impossible to sell.

***

In our time we know well the tedium that can be industrial life. Living in an ugly place full of factories. Working shifts at all sorts of ungodly hours. Doing a dull job that repeats over and over again. Having to inhale pollution every day and survive the absence of nature. More than anything, feeling like a minor and insignificant cog in a vast, mechanistic society. In the last years of the eighteenth century in England, where the industrial revolution was so visible, this sense must have been very strong.

The Romantics aimed to swing the pendulum back the other way.

Away from objective truth, societal 'progress' and cold rationality to subjective emotion, rapture, artistic freedom, and a re-unification of the individual with nature.[13] The notion of the artist alone in nature, gaining creative inspiration from awe-inspiring encounters with its force, wildness and beauty, once again became popular.

In his 1798 poem *The Prelude*, William Wordsworth writes:

> Wisdom and Spirit of the universe!
> Thou Soul that art the eternity of thought
> That givest to forms and images a breath
> And everlasting motion, not in vain
> By day or star-light thus from my first dawn
> Of childhood didst thou intertwine for me
> The passions that build up our human soul;

Not with the mean and vulgar works of man,
But with high objects, with enduring things –
With life and nature – purifying thus
The elements of feeling and of thought,
And sanctifying, by such discipline,
Both pain and fear, until we recognise
A grandeur in the beatings of the heart.
Nor was this fellowship vouchsafed to me
With stinted kindness. In November days,
When vapours rolling down the valley made
A lonely scene more lonesome, among woods,
At noon and 'mid the calm of summer nights,
When, by the margin of the trembling lake,
Beneath the gloomy hills homeward I went
In solitude, such intercourse was mine;
Mine was it in the fields both day and night,
And by the waters, all the summer long.[14]

It's got heart and spirit and mojo. It's life in hot blood – as Van Gogh wanted it. It embraces the world for what it is – rather than trying to 'progress' away from it. The same, romantic sentiments were expressed somewhat later by Walt Whitman in his '*When I Heard the Learn'd Astronomer*':

When I heard the learn'd astronomer,
When the proofs, the figures, were ranged in columns before me,
When I was shown the charts and diagrams, to add, divide, and
measure them,
When I sitting heard the astronomer where he lectured with
much applause in the lecture-room,
How soon unaccountable I became tired and sick,
Till rising and gliding out I wander'd off by myself,
In the mystical moist night-air, and from time to time,
Look'd up in perfect silence at the stars.[15]

What the Romantics were saying is "Hey guys...what about beauty? And rapture? And meaning? What about nature? Why don't we design a civilization in which we actually live, not just exist?"

If Wordsworth were in Julia's shoes, he'd have approached the other five people differently. Much more lyrically. He would have spoken to the heart.

Juan-Franco too, believes in this approach and when the time comes, will use it.

***

The industrial revolution was settling into the new normal. The usual, industrial way of being. 'Truth speaks for itself' had been widely accepted, but never tested amongst men of science. In ancient times it was the weaker but cleverer speaking to ancient man with the big biceps. Then it was Socrates speaking to the people of the city. Later, on science speaking to the clergy who believed that the universe revolved around them. All these ended the same way – badly. Perhaps, if a logical scientist spoke to other logical scientists, perfect input would lead to perfect output and a perfect address to perfect persuasion. By the middle of the nineteenth century, such an opportunity arose.

Let's assume hundreds of patients are dying because doctors have not yet figured out that they need to wash their hands between dissecting fresh corpses and assisting with childbirth.

A brilliant man named Ignaz suddenly realises what is happening and once he implements his theory, mortalities drop dramatically. When his suggestions are ignored, fatalities rise again.

In the world where rationality rules, Ignaz is a hero who wins the Nobel prize for medicine. In one where scientists as listeners are unable to leave their prejudice and emotions behind, Ignaz is ignored, marginalised for suggesting that gentlemen doctors can have dirty hands, beaten up and left in an asylum to die.

Ignaz Semmelweiz was a Hungarian with a PhD in medicine. He was appointed at the Vienna General Hospital in the Obstetrics Clinic, and was soon faced with an almost unbelievable riddle; there were two clinics where babies were delivered, but in the one the average mortality rate of mothers due to puerperal fever was about 10% while in the second it was only about 4%. Even more puzzling, for women giving birth in the streets, the mortality rate if anything, was lower than in either clinic. How could this be possible?

The year was 1847, and it was another two decades before Lister, prompted by the discoveries of Pasteur, would discover the existence of germs.

"It just can't be right" – or so thought Semmelweiz. In his adopted language: "*Das kann nicht wahr sein.*" He had to be missing something. Was the one clinic more crowded than the other? Yes, the second clinic was more crowded, because women dreaded going to the first. So despite being more crowded, the second clinic yielded fewer cases of puerperal fever. Was it the climate? No, the clinics were close to each other and the climate was the same. Could it be religious practices that were different? There was no evidence that this was the case.

The only difference was that Clinic 1 was a training facility for doctors while Clinic 2 was a training facility for midwives. Initially this didn't seem to make any difference either.

Then, through unfortunate circumstances and literally by accident, Semmelweiz's answer emerged. One day, his friend Jakob Kolletschka was performing a post-mortem and was pricked by a scalpel belonging to one of the students. He died soon afterwards of symptoms very similar to puerperal fever.

The penny dropped; the medical students were bringing death to the pregnant mothers.

This was Semmelweiz's realization. It wasn't the scalpels, because the unfortunate women who died were not pricked by scalpels. It was the students' hands. In Clinic 1, the medical students would do autopsies on cadavers and then go straight over to deliver babies. In Clinic 2 there were no medical students – only midwives, who did not do autopsies.

The medical students were carrying something on their hands from the autopsies. Whatever this was, it was killing many of the delivering women. And the answer was simple – they only had to wash their hands in between, and the problem would be solved!

Immediately after Semmelweiz instituted these changes, there was a dramatic change: Clinic 1's mortality rate dropped from 18.3% in the previous month to 2.2% two months later, and later to zero for two full months. It was measurable and, if required, could be repeated by others. It satisfied the requirements of the scientific method.

Later, Semmelweiz realised that it wasn't only exposure to cadavers that could have this adverse effect. Together with his colleagues he had examined a woman in labour with a severe cervical cancer and then the doctors (without washing their hands) had delivered twelve babies from other mothers. Eleven of these women died.

The problem wasn't just cadavers, it was 'putrid living matter' as well. But simple hand-washing was the solution.

In the Hollywood version the entire world would soon have been converted to the simple discipline of hand-washing before moving from cadaver dissection to assisting women in childbirth. Thousands of lives saved. But no – Semmelweiz was not a good politician. He aggravated people. He seems to have been very direct in his approach. He disliked the career medical bureaucrats. Moreover, he was a foreigner, and an ill political wind was blowing – 1848 has been called the 'Year of Revolution'.

But these were perhaps still peripheral issues: Far more important for present purposes was *what it would mean* if Semmelweiz was proved right. This would mean that

- the highly esteemed gentlemen doctors had hands so dirty that you might die if they assisted you during childbirth, and that
- they were so ignorant that they did not even know this. Such,
- much of accepted medical doctrine had to be wrong.

There are some truths that just cannot be believed, and these were among them.

One comment from a respected medical expert was "... with due respect for the cleanliness of the Viennese students, it seems improbable that enough infective matter or vapour could be secluded around the fingernails to kill a patient".[16]

Semmelweiz's ideas were dismissed and his career path repeatedly blocked. Year upon year his frustration grew. His successor in Vienna published a textbook with 30 causes of puerperal fever – the link with cadavers came in at number 28. Seven years after his discovery, 400 women died of the disease in Vienna alone. His theory was even rejected by Rudolf Virchow, the foremost authority of the time.

By 1861, it had been well more than a decade and there was still no sign that his theory would ever be accepted. Semmelweiz descended into severe depression, became extremely confrontational, and was increasingly isolated. In 1865, a colleague lured him to an insane asylum under false pretences. Semmelweis tried to leave but was cruelly beaten by several guards, put in a straitjacket and thrown in a dark cell. He died soon afterwards from blood poisoning and with extensive internal injuries.[17]

The messenger was dead.

Just two years later, Lister was well on his way to proving the existence of germs. In time it was realized that the dead messenger had been correct.

But it was twenty years later, and a new generation of doctors was out there. Of them, fewer had as yet failed to wash their hands after a dissection of a cadaver and immediately thereafter assisted in childbirth – then lost the patient. The younger doctors' hands were clean, and they could admit that Semmelweiz had been right without simultaneously admitting that they'd been directly responsible for their patients' deaths.

Humans are only occasionally rational, as an investigation of the research will show. But believing the beautiful lie had cost Ignaz Semmelweiz everything.

Truth just wouldn't speak for itself.

***

Soon, Esthia will walk into her office, half a double espresso in hand, and switch on her PC. When she sees news of the cyclone, she will quickly search for visuals and see the stunning satellite image – white, huge, circular and voracious.

To her, a scientist, there wouldn't be a shadow of doubt that a cyclone named Fifi exists. She would be looking at it. And yet, on the yacht, Erin is drinking her Mochito. Sam is lying on his back, face towards the sun, eyes invisible behind his sunglasses. Julia is inside, trying to see whether she can restore reception.

Julia and Juan-Franco believe what Esthia will soon know to be true, but Sam and the other three don't believe.

A cyclone is coming. It's true. But that truth by itself has been spoken. Julia has rushed out and told everyone, and failed to convince the majority of people on the boat. Despite mounting evidence, it will continue to be 'not-believed'.

The truth isn't speaking for itself. Isn't convincing by itself. Truth needs a ribbon in her hair.

3

ESTHIA

Esthia gets to her office at a quarter to nine. She hangs her handbag over her chair and puts down what is left of her coffee on the table. Towards the mouse pad, on the right. Five minutes ago, she stood at a coffee vendor outside. It's a cold day – mid November. Two days back the first real snow of the season came floating down, unobtrusive as a whisper. This morning, when she kissed her husband goodbye and walked out the door, she could feel it – the unobtrusiveness had become a presence. Early mornings and late afternoons as she walked.

She always walks through the park, no matter what the weather. She takes an umbrella now that's she's a little older. In winter she wears a waterproof coat.

Some years ago, when her English was still imperfect, she once asked Juan-Franco if he wanted to walk "under the rain" with her. It was the first thing they did together. In another park, another city. They walked, past the places where the would-be ballerinas balance on pillars for small change as they practise their moves, and where the people queue for opera tickets at night. Next to the road were coffee shops and places selling music and musical instruments. Memorabilia from the summer of love. T-shirts with Dylan and Che Guevara.

At first the rain was light, then much heavier. She could see Juan-Franco looking at her for a reaction, but she didn't accelerate, didn't pull her head down, didn't take out an umbrella, didn't hold an old, folded newspaper over her head. Esthia just walked, chin up, and spoke about songs she liked. About her country and growing up. The repression and abuse and its scars. About Pablo Neruda.

As he watched her, his slumped, rain-repelling shoulders straightened like hers. His chin lifted, like hers. The rain drops were no longer hitting him at the back of his head, but on his forehead and cheeks. He told her about the guitar he built as a youngster from a book, about the songs he was writing, about the unbalanced wheel of life. Businessmen and beggars. How each has what the other craves.

They arrived drenched in ways that would never dry and went for a cappuccino.

Esthia smiles as she switches on her PC. It's her fondest memory, along with the time they saw the North Star brightest in a very starry sky and he sang an entire Don Mclean song to her *a capela.* It was about Van Gogh, and the last verse always rang in her head when she saw the North Star:

> And when no hope was left in sight
> On that starry, starry night,
> You took your life as lovers often do,
> But I could have told you, Vincent...
> This world was never made for one as
> beautiful as you.

At the coffee vendor six minutes before, she asked for a double espresso. As she stood waiting, the vapour from the tube station below enveloped her, and she recalled all those old movies she used to watch with Juan-Franco. The heroine on the platform, disappearing in the steam of the locomotive. Waiting for the home-coming hero.

Her thoughts slip to the day ahead. Writing up her recent research that she'll be doing. *Colony Collapse Disorder.* The bee colonies are collapsing – nobody knows why. But she's identified a possible link with mobile phone towers. An academic journal has asked Esthia to submit an article and she'll be spending the day writing.

Below in the street, a motorist hoots urgently. Esthia freezes, remembering the ferries in the Greek Islands sounding their horns into misty mornings. The time near Samos, Juan-Franco already awake as she opens her eyes – smoking and watching her.

It was that afternoon that she told him they couldn't be together. Not in the flesh. That it was too much, too perfect, that she couldn't handle it. His eyes aged as she spoke, but retained the gentleness. To this day, when once in a long while they go for a coffee, it's still there.

Even now, when she reads his emails or listens to the demo CDs he sends her, she sees that gentleness. And behind him, Samos and the Aegean Sea. It's everywhere that day as he watches her wake up on the ferry.

Esthia takes another sip of her double espresso. A small one. She's past halfway, and the taste in the hastily stirred styrofoam cup is progressing from strong to sweet as she nears the bottom.

After that day in Greece, she'd never had to talk about such hard things again. Never had to recall her childhood again. She thinks about her academic article and the fact that she doesn't have to talk to the people who want to know why the bees are dying. As she walks to the stationery desk to fetch printing paper, she has a spring in her step. She's only talking to the editor. She finishes her espresso and turns her attention to her PC, which is now online.

Esthia sees the storm warning in the general newspapers, moves quickly to the weather channel, opens the other-worldly graphic of what it looks like in all its monstrosity, immediately sees the convergence of its projected course with where the yacht must be.

As she sits down heavily, she thinks for the first time in years of Pablo Neruda.

4

THE ENDGAME OF BEAUTIFUL LIES – WARNINGS AS FICTION

Let's imagine a world where nothing is real. Nothing is what it appears to be. Fire might be water. Earth might be wind. The quintessence of everything is that it appears to be something else. Appears to be what it's not. Whenever we reach to touch something, we touch something else. We put a hand in the wash basin to wash it and get burned by fire. We try to bury a friend but find ourselves digging a grave in the wind.

All that we know about everything is how it appears to be. And we know that what it appears to be, it is not.

In this world, nobody can be convinced of anything, because by telling him that it is green, he knows it cannot be. By telling her it is hot, all she knows is that it's not.

In this world, people enjoy the way things appear. Those that appear to be the most splendid are enjoyed the most. What these things really are, matters not. Nobody tries to find out, for there is no way to do so. When everything is a lie, everything is true.

And so people just look at what they know not to be; to not be what they see, and enjoy it for what it isn't. But claims to be.

This is the endgame of beautiful lies. Nero's world. A place where there are so many lies that the concept of truth has faded away, and with it the concept of the lie. Leaving only beauty and ugliness, the compelling and the commonplace.

In this world, everything is fiction. And all warnings are fiction.

In this world, a radio report about an impending storm is no different from a novel about an American version of Bridget Jones in New York.

It's all fiction. Amusing or not, depending on its beauty. This is the world that we're approaching today. Let's see how we got here and what it means for Julia and Juan-Franco, and for Esthia. And let's allow Julia to travel through time and space, master the languages and dialects wherever she goes, and utter her warning at different times and places in history, to see how it is received.

***

It's summer in ancient Greece, the kind of white, scorching day Socrates would experience in captivity in the last month before he died. The shepherd boy is on a hilltop, where the lightest of breezes is stirring. Below and around him are his sheep, listless in the heat, showing no inclination to wander away.

"Good day madam", he says to the woman who has so mysteriously appeared next to him. "Who are you?"

"I'm Julia", she replies, smiling at his large, dark eyes and mischievous mouth. "How's your day going?"

"Boring", he asserts. "So boring. These sheep are too tired to go anywhere. Even a wolf won't show himself in this sun.". He clasps his staff nevertheless – if the wolf ever comes, he'll be ready.

"I could have stayed in the village or gone swimming with my friends", he continues. "There's nothing for me to do here. My sheep don't need me until later. In the late afternoon when it cools down, some will want to wander off and I can catch them. And a wolf may even come. But for now I am wasting my time."

Julia nods and sits down on a stone next to him. It's so hot it almost burns her. He asks what she does. "Today, I bring a warning", she says. "Bad weather. A big storm. It is my job to warn people."

In one world, the shepherd boy offers to walk down with her to the village. He takes his sheep with him. "If there's bad weather coming, my sheep must be safe in the pen", he says.

Down in the village, people are sleeping, waiting for the heat to ease. The shepherd boy goes from door to door and calls everyone out with scant regard for their grumpiness. "Quick, quick," he urges, " there is a lady who has come to warn us of a storm."

Soon they are all in the town square looking quizzically at her clothing.

"There's a storm coming," Julia says, "you have to keep the sheep in their enclosure Keep enough food and water in your houses, bring in your children. There will be a very strong wind."

“Who are you?”, a dark haired woman asks, “How do you know this?”

Julia thinks for a moment. “I come from afar”, she says. “Someone came to tell me he stood on a mountain and saw an immense dust cloud moving this way. And the sky was very dark”.

The villagers believe her, thank her, invite her to take shelter with them.

In a second world, the shepherd boy doesn’t offer to walk down with his Julia. His mischievous eyes light up, the boredom is gone from his posture. He’s all alert, all attention. “A warning...”, he starts, thinking, “...mmm...now I have a plan!”

He gets to his feet with purpose. “A plan for this boring day!” He skips down towards the village. As he goes he shouts as loud as he can, with compelling urgency in his voice: “WOLF! WOLF! THE WOLF IS CHASING THE SHEEP!”

The villagers immediately awake from their slumber and come running up the hill to help the boy drive the wolf away. But when they arrive at the top of the hill, there is no wolf. The boy laughs at the sight of their angry faces.

“Don’t cry ‘wolf’, shepherd boy!” say the villagers, “when there’s no wolf!” They go grumbling back down the hill.

Soon after, the boy sings out again, “WOLF! WOLF! THE WOLF IS CHASING THE SHEEP!”

To his naughty delight, he watches the villagers run up the hill to help him drive the wolf away. Because a wolf is a very serious matter.

When the villagers see no wolf they sternly say, “Save your frightened song for when there is really something wrong! Don’t cry ‘wolf’ when there is NO wolf!”

But the boy just grins and watches them go grumbling down the hill once more.

In this second world, Julia is still sitting on the hot stone with her strange clothing. The villagers see her but they are worried about a wolf – and don’t take the time to engage her. The shepherd boy is still full of adrenaline and mischief, but he realizes that Julia is still waiting.

“Let me walk down to the village with you”, he says to her, “so you can do your work as the one who has to deliver a warning. I’ll take my sheep with me – if there’s bad weather coming, they must be safe in their pen.”

But Julia shakes her head, “You’ve ruined it for me, shepherd boy”, she says sadly. “They’ll never believe me now. Twice you’ve woken them to cry “WOLF!” They are angry and distrustful now and will only believe someone with a warning again after many days. If the storm comes now, they’ll be unprepared, and there’s nothing I can do about it.”

Just then, as the sun dips lower and the heat begins to dissipate, a wolf appears and starts chasing the sheep. Alarmed, the shepherd boy leaps to his feet and sings out as loudly as he can, "WOLF! WOLF!"

But the villagers think he's trying to fool them again, and so they don't come.

At the sun sets, everyone in the village wonders why the shepherd boy hasn't returned to the village with their sheep. They go up the hill to find the boy. He's weeping. As Julia prepares to leave for another place and another century, she hears him saying:

"There really was a wolf here! The flock has scattered! I cried out, 'Wolf!' Why didn't you come?"

An old man tries to comfort the boy as they walk back to the village.

"We'll help you look for the lost sheep in the morning," he says, putting his arm around the youth, "Nobody believes a liar, even when he is telling the truth".[18]

***

It's a dark night in Nottingham in March 1811. People have been looking for signs of spring, as they do at the very end of winter. Is there a balmy breeze pushing through their hair yet? Are the blossoms perhaps visible on the trees? But no, the cold, dead season insists on completing its full tenure. Perhaps another four weeks, perhaps another six.

Now, in the small hours, it's nippy and Julia isn't dressed for it. She remembers the heat in Greece and the hot stone she sat on. Her hands go to the opposite shoulders, embracing herself and crossing her arms in front of her body. Not only for the cold – also because she's afraid of the three startled men with the sledge hammers, in front of her.

They're inside the textile factory and a single lantern provides the only light, creating large and grotesque shadows around the four people facing one another. The three Luddites entered the factory through the back door. Their leader, John Berry, worked here once, and still has a key. On their way out, they will also smash the door so the suspicion does not fall on him – but they need to maintain silence and secrecy for as long as possible. If they get caught, they're likely to be tried and hanged.

"Who are you?", hisses Berry at Julia in a low voice, bringing the sledge hammer to his shoulder. "You shouldn't be here..."

His cohorts likewise look threatening and suspicious. But at least Julia's light and alien clothing doesn't place her in any immediately discernible social class. She tries to react quickly, before they can take action, but calmly, so she doesn't incite them any further.

"I'm from far away", she says. "I bring a warning about a storm."

Her accent confirms her words and she sees one of them cocking his head to the side as he grapples with it. "A storm?" Berry asks, "what kind of storm?"

"A hurricane with lots of rain", says Julia.

"Who are you? How do you know this?" asks the one who is struggling with her accent.

"I'm from far away", she repeats. "But I heard a weather report over the radio. It was a storm warning. You should tell your families and bring everyone inside and batten down the hatches."

"What is a radio?" asks Berry. Julia sees him lowering his sledge hammer and the tension leaving his shoulders. He still speaks softly though, as she's doing, and looks over his shoulder towards the open back door from time to time. "And what is a weather forecast?"

The questions have come unexpectedly, about basic things. Important questions. Being in a factory, in the midst of technology, Julia decides that it's possible to relate the answer to them in a way they can follow. "A radio is a machine made so people can hear others talking from a great distance. A weather forecast is a communication about the weather that experts make after using machines to determine what will happen in the next day or two. I heard such a report while listening to the radio. The man speaking was far away but his instruments showed him there is a hurricane coming."

There is silence. The lantern flickers, the shadows take on new and eerie forms as the four people move. Outside, audible through the open door, there is a breeze, and the draft makes Julia shiver.

"We don't like machines", Berry starts. "All our families were lacemakers, for generations. That's what we did. And now these machines are doing it instead ...!"

He turns towards the machine closest to him. His grip has firmed on the hammer, the tension is back in his shoulders.

"It's war for us, and the rules of fair play do not apply in love and war. We've come to smash the machines."

Julia looks at them, three large and muscled men, and feels pity. When she leaves, they'll wield their hammers and within minutes very little will be left of the factory. Their brute force will win the battle but lose the war. For new machines will come regardless, and even if these three are not apprehended and brought before the magistrate, their families will never make lace again. Their war is lost.

"I have to go now," says Julia. "there are people on a boat who need to hear my warning. When you get home, tell your families, there's a hurricane coming."

She leaves through the open door, the one that will be smashed soon. Behind her, having just heard about more, new machines that will be doing even more things that tend to make humans obsolete, there is even greater anger in the three men than before. They too sense that their war is lost, and the rage and frustration find expression in the speed and the power with which they swing their sledge hammers. Julia hears metal on metal, glass shattering, a rising cacophony of helpless resistance to the inexorable course of history. But she knows that when they get home, they will warn her families about the hurricane. They have never seen a radio nor the instruments used to compile a weather report, but they have seen machines more generally. Seen their awesome power and reliability and how they totally transform the world these men were born into. They may hate machines, but they respect them, even trust them. And so when Julia brings them a warning based on machines – even ones they have never seen – they believe her.

***

The Luddites would continue their resistance, clashing with the army and the law. Some were executed and others shipped off the penal colonies, never to be seen again. Within a few decades, their fight was over. New machines continued to come. Weather reports and radio within the same century, along with electricity. Television a little later. Within another century; internet, smart phones, text messaging, Twitter, Facebook, Instagram.

Even when it was just radio, an irresponsible shepherd boy could suddenly, if he wanted to, trick not just one village of people but an entire city – or even large parts of a country. It would take forty years after the invention of the new medium before that shepherd boy would come along. His name would be Orson Welles, and his legacy would, some years later, contribute to making Sam lie down on the deck of the yacht just after hearing a cyclone warning, confident that no storm was approaching.

It's October 1938 and Julia is in New York, at CBS radio, in the studio with the precocious radio actor.

"Good day, madam," he says to the woman who has so mysteriously appeared in his studio, "who are you?"

"I'm Julia", she replies. "How's your day going?"

"I'm Orson. My day's exciting, very exciting", the handsome, dark actor replies. "We're broadcasting a hoax just now, with it being Halloween and all. A drama called *The War of the Worlds*, you may know the book. About Martians invading the earth. We want to make it sound like it's really happening. We'll use news-style reporting and frequently interrupt the programme with 'newsflashes'. What brings you here?"

"Oh, I'm a scientist", she replies. "I have to warn people about a coming storm. A real twister."

"Mmm", Welles nods. "I'm sure someone will interview you just after the show. Good if folks are forewarned. Just hang around, it'll be an hour or so."

The rest of the cast starts arriving. Julia reads the newspapers from the last weeks, lying around, and soon gets the impression of a country and a world on edge. It hasn't been just another fall.

In Asia, Japan is at war with China. In the heart of Europe, Germany is preparing to march to the incantations of a raving Adolf Hitler. A month before, the Fűhrer's vitriol was heard around the world, also on radio. Angry demands were made. Boots of marching men were lifted high and in unison, and brought down hard in unison, as if to portend what was to happen to anyone who got in the way.

In Munich, the international community has just offered the Fűhrer a giant present in the vain hope that his territorial hunger will be permanently stilled. Sudetenland has been cut out of Czechoslovakia, like a heart from a living organ donor, and given to Germany. The military alliance between Czechoslovakia, England and France is in in tatters, the little guy sacrificed to appease the big guy next door.

As the leaves started to fall from trees in France, in England, in the USA, in Holland and Belgium, it seemed prophetic of other things that were to follow – to fall. People were afraid. They felt unsafe, didn't know what might happen next, but they knew whatever it was, it might be really bad.

Julia looks up again and sees the orchestra arriving for the show. Their leader is called Bernard and they get ready to play. The actors too, are there. Nobody minds her where she sits – everyone assumes she's there for a reason. Part of the show in some way, of what would soon be called 'the greatest hoax of all time'.[19]

Welles calls everyone to silence. "The legal department has forced more than twenty script changes on us", he informs them. "We've had to change the names of real people and places to avoid possible legal action – but we kept them as close to reality as we could. Other than that, it's as we've rehearsed it. We'll announce that it's fiction three times – the first at the twelve minute mark, when most folks will still be tuned in to *Chase and Sanborn Hour* on NBC. They won't know – but will switch to us in time for our restart and the climax of our story. That gives us another 28 minutes before the next ad break, which will only occur after the climax of the story. We'll state again that it's fiction at the very end, when it won't matter much anymore, anyway".

As the start draws near, a sense of gravity descends, as Julia would expect when performing a live radio drama. There's no space for mistakes. They had to capture their audience's attention right at the beginning and keep it throughout.

Then they are live, and Welles starts, playing Professor Pierson:[20]

> "We know now that in the early years of the 20th century this world was being watched closely by intelligences greater than man's and yet as mortal as his own. We know now that as human beings busied themselves about their various concerns they were scrutinized and studied, perhaps almost as narrowly as a man with a microscope might scrutinize the transient creatures that swarm and multiply in a drop of water. With infinite complacence people went to and fro over the earth about their little affairs, serene in the assurance of their dominion over this small spinning fragment of solar driftwood which by chance or design man has inherited out of the dark mystery of Time and Space. Yet across an immense ethereal gulf, minds that to our minds as ours are to the beasts in the jungle, intellects vast, cool and unsympathetic, regarded this earth with envious eyes and slowly and surely drew their plans against us. In the 39th year of the twentieth century came the great disillusionment..."

Three of the actors will play announcers or radio reporters reporting from the field. The most prominent, Frank Readick, is playing a radio announcer, called Carl Phillips. For most of the week he's been in the record library, listening to Herbert Morrison's radio report of the Hindenberg disaster. The tone as the airship crashed and burnt, the utter sense of immediacy, the excitement mixed with horror. These were the subtleties he would be looking to recreate as Welles breaks to what is styled as a weather report, and introduces studio announcer Dan Seymour to deliver it. In fact, Welles is intending to get the listeners accustomed to breaks in the programme that later will appear to be the result of the chaos of the Martian invasion, causing scheduled programming to cease. Welles and his crew are supposed to be performing a drama, but they are pretending to present quite drab radio on a quiet Sunday evening. All the pegs in the ground of reality are being moved or removed. People tune in to listen to a drama but are informed about the mundane things that occur in everyday life. A weather report. Spanish music. They're being told the drama isn't a drama, it's a boring night at CBS filled with this and that.

> "For the next 24 hours not much change in temperature...", the studio announcer begins. "A slight atmospheric disturbance of undetermined origin is reported over Nova Scotia, causing a low pressure area to move down rather rapidly over the northeastern states, bringing a forecast of rain, accompanied by winds of light gale force. Maximum temperature 66; minimum 48. This weather report comes to you from the Government Weather Bureau. ... We now take you to the Meridian Room in the Hotel Park Plaza in downtown New York, where you will be entertained by the music of Ramon Raquello and his orchestra."

The in-studio band starts playing but the fun is about to start. Julia can sense the mischief in the entire cast. The music is abruptly interrupted as the second announcer enters the fray:

> "Ladies and gentlemen, we interrupt our programme of dance music to bring you a special bulletin from the Intercontinental Radio News. At twenty minutes before eight, central time, Professor Farrell of the Mount Jennings Observatory, Chicago, Illinois, reports observing several explosions of incandescent gas, occurring at regular intervals on the planet Mars. The spectroscope indicates the gas to be hydrogen and moving towards the earth with enormous velocity. Professor Pierson of the Observatory at Princeton confirms Farrell's observation, and describes the phenomenon as 'like a jet of blue flame shot from a gun'. We now return you to the music of Ramon Raquello, playing for you in the Meridian Room of the Park Plaza Hotel, situated in downtown New York..."

Julia watches and listens as the band plays again. She catches Welles' eye. He's absolutely buzzing. Again there's an interruption in the music – CBS tells the audience that they'll now move to Princeton to interview a famous professor on what is happening on Mars. In dry and scientific language, Welles himself as interviewed by Readick, says that there is almost certainly no life on Mars and not much to report. But then someone hands him a note, while the interview is live – a cable about a seismic shock near Princeton. Still maintaining his scientific detachment, Welles dismisses any link with Mars and says it must be a meteorite that has fallen.

The show moves back to the studio, where the announcer quotes various news reports from other geographies, reporting odd happenings around Mars and the skies of planet earth. Julia looks at the band members and sees the gravity of the illusion starting to hijack their demeanour. No longer the smiles and twinkling eyes. For them, on some level, this is happening. Mars is invading earth, but the dots are only slowly being joined. In millions of homes, the same is happening, and soon chaos would engulf America – not the chaos depicted in the show, but the chaos caused by the show.

As far away as in Canada, people fleeing their homes and clogging the highways. A coincidental power failure in Washington, believed to have been caused by the invaders and further feeding the frenzy. Distressed callers phoning CBS's affiliate company VGAR in Cleveland and being reassured that everything is okay. Some responding by alleging that CBS is covering up the truth and that the Martians have indeed landed. In the confusion, other listeners decide that it is Germany that's invading the USA.

But here, in the studio, there is music again. Swing, soon interrupted to go to the farm where the thing has landed. The professor and the interviewer have raced there and are on air:

> "Ladies and gentlemen, this is Carl Phillips again, at the Wilmuth farm, Grovers Mill, New Jersey. Professor Pierson and myself made the eleven miles from Princeton in ten minutes. Well, I ... I hardly know where to begin, to paint for you a word picture of the strange scene before my eyes, like something out of a modern *Arabian Nights*. Well, I just got here. I haven't had a chance to look around yet. I guess that's it. Yes, I guess that's the ... thing, directly in front of me, half buried in a vast pit. Must have struck with terrific force. The ground is covered with splinters of a tree it must have struck on its way down. What I can see of the ... object itself doesn't look very much like a meteor, at least not the meteors I've seen. It looks more like a huge cylinder."

There's an interview with the shaken farm owner who saw it landing. The descriptions sound more high tech than like a meteor. A spacecraft maybe. And then, horror upon horror, the top lifts and something starts to emerge as an overcome Phillips keeps the listeners apprised:

> "Good heavens, something's wriggling out of the shadow like a gray snake. Now it's another one, and another. They look like tentacles to me. There, I can see the thing's body. It's large, large as a bear and it glistens like wet leather. But that face, it ... Ladies and gentlemen, it's indescribable. I can hardly force myself to keep looking at it. The eyes are black and gleam like a serpent. The mouth is V-shaped with saliva dripping from its rimless lips that seem to quiver and pulsate. The monster or whatever it is can hardly move. It seems weighed down by ... possibly gravity or something. The thing's raising up. The crowd falls back now. They've seen plenty. This is the most extraordinary experience. I can't find words ... I'll pull this microphone with me as I talk. I'll have to stop the description until I can take a new position. Hold on, will you please, I'll be right back in a minute... A humped shape is rising out of the pit. I can make out a small beam of light against a mirror. What's that? There's a jet of flame springing from the mirror, and it leaps right at the advancing men. It strikes them head on! Good Lord, they're turning into flame!. Now the whole field's caught fire. The woods ...the barns ... the gas tanks of automobiles ... it's spreading everywhere. It's coming this way. About 20 yards to my right ..."

Julia listens to the field report telling how all the bystanders are obliterated by heat-waves. The reporter's panicked cries about the oncoming flames disappear abruptly from the air.

From there, Welles stages chaos inside the radio station to underline the seriousness of the situation. Programming is erratic and frequently interrupted with worrisome introductions like: "We interrupt this programme..." There is a report of soldiers sent to attack the Martians:

> "Here is a bulletin from Basking Ridge, New Jersey: Coon hunters have stumbled on a second cylinder similar to the first embedded in the great swamp twenty miles south of Morristown. Army fieldpieces are proceeding from Newark to blow up second invading unit before cylinder can be opened and the fighting machine rigged. They are taking up a position in the foothills of the Watchung Mountains. Another bulletin comes in from Langham Field, Virginia: Scouting planes report enemy machines, now three in number, increasing speed northward, kicking over houses and trees in their evident haste to form a conjunction with their allies south of Morristown. Machines

also sighted by telephone operator east of Middlesex within ten miles of Plainfield. Here's a bulletin from Winston Field, Long Island: Fleet of army bombers carrying heavy explosives flying north in pursuit of enemy. Scouting planes act as guides. They keep speeding enemy in sight. Just a moment please. Ladies and gentlemen, we've run special wires to the artillery line in adjacent villages to give you direct reports in the zone of the advancing enemy. First we take you to the battery of the 22nd Field Artillery, located in the Watchtung Mountains..."

But the soldiers are mown down by the Martians, to a man.

CBS continues to tell the nation how it's being invaded by aliens. There are breathless updates of casualties and the utter destruction of infrastructure by the Martians. A politician imitating the president's voice talks to the nation.

In the studio, producer John Houseman receives a phone call ordering him to interrupt the transmission and reassure everyone that it's just fiction. But the break is just a minute away, when such an announcement will happen anyway. He waits. At the switchboard, thousands of people are trying to get through telephonically, all at once. Then the press starts arriving. In the weeks to follow, newspapers would publish a deluge of stories about the hoax and there would be calls for the introduction of radio censorship to prevent a recurrence. The message would be carried to every nook and cranny, saying essentially: "Orson Welles at CBS has used fiction dressed up like news to bamboozle the audience – and he pulled it off."

Back on air, updates from the 'front' continue, the listeners hear that the Martians are now releasing poisonous gas – again a reporter is overcome and after a fit of coughing, is heard no more. More spaceships are landing all over the country and the epicentre of events soon moves to New York City, where giant Martian machines are entering the city with people scrambling away, dying, running, "diving into the East River like rats". At the end of this, the second episode, a lone-voiced radio operator is calling CBS, asking repeatedly and plaintively: "Isn't there anyone on the air..?"

The silence affirms that even the radio station is gone, along with the rest of the city.

There is then a break at about the forty-minute mark, with a reminder that the programme is fiction. Julia sees policemen coming into the building, first a few, then many more. As they go back live the story winds down, with the Martians succumbing to earthly germs that had not been expected and against which they had no immunity.

Then, at the end, follows the third disclaimer. Welles tells the audience that it was a Halloween show. It was a scare tactic, a way of shouting "Boo!"

Looking at the police and media and company lawyers all over the place, Julia sees Welles drop his head in his hands and start to mutter to himself. He feels that his career has been ruined, doesn't know yet that he's just turned himself into a lifelong celebrity who would soon make one the best films of all time, using very similar tactics.

She recalls telling the shepherd boy that he's ruined it for her, that nobody would believe her now. She cannot help grimacing at the realization of how those words completely robbed of her of a vocabulary to describe what Welles has just done. A deception, far, far greater, at a scale unimaginably larger.

Right now, she couldn't possibly get airtime to address anyone about any threat, no matter how grave.

Julia returns from her travels to the yacht. With her just having been part of a thought experiment on how her warning would be received at different times and places in history, she of course remembers nothing of her incredible journey through space and time and approaches her task on the yacht just like before. As a scientist. Like a scientist.

Around her, the world we saw evolving from the shepherd boy to the Luddites, to Orson Welles, continues to change.

***

If we look around us today, everywhere there is fiction being packaged as reality.

As a first example there is the mega-world of professional wrestling brought to us via television. The company WWE, trading in wrestling events and sports entertainment, is listed on the New York Stock Exchange. In 2012, it had headline revenue of USD 600 million and profit of USD 40 million. Its total assets are in the USD 400 million range. What is marketed to us isn't just wrestling bouts, but theatre. There is bragging, intimidation, promises, deceit, cheating, scheming, back-stabbing. There are cameras backstage, capturing the drama even when three or four people are having a private conversation or when one sly wrestler is planning to surprise another somewhere outside the ring with an attack. There are the clever speeches made by contestants before fights, talking directly to the fans in the stadium, playing to every pseudo-psychological theme conceivable.

There are six championship belts being fought for, amongst others the

World Heavyweight Championship. And between all of this we are asked to believe that this is for real, that these guys fight to win, that nothing is scripted and that show-business plays no role in the outcome. We are told that these guys fight to win and that is fact.

There is the issue of banking, where the bank you or I have chosen may have no bricks and mortar whatsoever. It might be in Iceland and have former fishermen working for it in key banking positions; might have a beautiful website and a confidence-boosting log-in procedure. It might also have a dismal deposits-to-debt ratio of 0.4:1 instead of the recommended 1.25:1. But we would happily bank with it – because internet-wise, it seems better than the others. We see virtual smoke and mirrors, but we believe it as if it is fact.[21] And alas, soon there is much smoke and flames plus many tears, but no more mirrors.

In the virtual world, we build towns and own property. We pay vast sums of money for virtual weapons so we can fight better in virtual games. When we pass away, the executors of our estates have to deal with the assets we have accumulated in the virtual world.

In the cinema, we believe a plot about someone who doesn't know that his entire life is set within a giant television studio and who is being watched by millions of viewers 'out there'. This happened in the *Truman Show*, starring Jim Carrey and Ed Harris and released in 1999.[22] Truman Burbank has been in front of the cameras since before he was born. All those around him are in on keeping him in the dark – actors who come to play in the show. Everyone knows that what is happening is a show – except Truman himself. During the movie he starts to figure out that something is amiss with his world – on the outside there is now a *Free Truman* campaign. At the end, Truman is nearly killed by the producer in a 'storm', but he manages to escape. In one of the last scenes, Truman's former audience is shown – searching for something else to watch. The film cost about USD 60 million to make and earned about USD 260 million at the box office.

At about the same time *The Matrix* was released.[23] In this film, the character played by Keanu Reeves is a hacker who (much like Truman) discovers that reality isn't what it seemed to be. Everything is simulated computer reality – this 'reality' is created by computers and machines to pacify humans, who are useful to the machines to harvest body heat and energy as a power source for themselves. From there, the rebellion against the machines (and against the fake reality) develops. The film cost USD 63 million and made 463 million. It was also followed by two sequels.

More recently there was the block-buster *Inception*.[24] Cobb (Leonardo Di

Caprio) takes the lead role. He makes a living by infiltrating the subconscious of espionage targets and planting ideas so as to extract information from them. This takes place through dreams, sometimes through dreams within dreams. In the real world, people are heavily sedated in order to allow these dreams to occur. In each level down, time passes more quickly – thus, in a dream within a dream, a matter of hours can be fifty years. It is revealed that Cobb and his deceased wife Mal (Marion Cotillard) had lived in such a state together, designed and built cities, and grown old together before she'd killed herself in the misapprehension that she was still dreaming.

Each person has a specific object that they carry with them to distinguish dreaming from reality – Cobb's is a spinning top.

Inception cost about USD 160 million to make and at last count had made over USD 800 million.

It is clear that the line between fiction and reality has become so blurred that multiple best-selling films can be made with this same, essential theme. Audiences have no trouble in suspending disbelief for a storyline that suggests that we might be completely wrong about everything we believe about the world and that our consciousness might be generated by computers, machines, or the director of a TV show.

***

We pay good money to watch 'live music'. About 45 years ago, that might have been Jimi Hendrix setting fire to his guitar at the Monterey Festival, then smashing it on the floor. That was pretty live. In Greenwich Village in the early sixties it might have been Bob Dylan playing guitar and harmonica at the same time to get the desired effect, and singing in between.

Today, we have a very different industry. Going to a concert by Janet Jackson or Madonna (or more recently, Britney Spears) is as close to gymnastics and the circus as it is to singing. It's a spectacle, yes, and the star attraction can dance, and his/her timing and rhythm is undoubtedly excellent because the "lip-synching" with the pre-recorded music is often nearly perfect. In an interesting irony, Spears' 2009 tour was billed "The Circus Starring Britney Spears."

We want perfect shows with toned artists hitting perfect notes even as they perform astonishing dances. Sometimes, on some songs, it just isn't possible, and the sound is pre-recorded while the artists mime or "lip-synch".

Other times, the guys doing the lip-synching aren't even the guys doing the singing. This happened with Milli Vanilli, who was unfortunately caught out and stripped of a prestigious music award.

Many other artists are known to have lip-synched even famous performances – like Luciano Pavarotti miming his last performance – a pre-recorded "Nessun Dorma" at the winter Olympics in Turin in 2006.[25]

"Live" doesn't necessarily mean we hear the singer sing anymore. It might mean we see the singer and you hear what they sound like in the studio.

In music video, we don't even have an expectation anymore that the artist is singing where the footage is being shot. Lately, producers have sometimes saved us from singers singing on the beach, in the desert, or while riding a horse. They shoot the footage without the singing and mix with other footage that shows the artist singing in a setting where singing is somewhat believable.

And yet, the hills of Austria are forever alive with the ubiquitous Sound of Music.

Then there is the world of reality TV, where almost nothing is real. Scenes are scripted, steered, and re-shot. It's show business, baby. It's just pretending to be real, that's all.

***

What does all this mean for Julia and Juan-Franco?

What world has permeated their minds and the minds of all the others on the boat? How should you and I, when our lives might depend on convincing others of a threat they cannot yet see, approach the matter?

As a civilization, we've thoroughly mixed up fiction and reality in the perception of the public. Journalist and author Frank Rich writes:

The boundary between reality and fiction has now been blurred to such an extent by show business, the news business and government alike that almost no shows produced by any of them are instantly accepted as truth. The market for fake news has become so oversaturated that a skeptical public is finally dismissing most of it as hooey until proven otherwise...[26]

"WOLF!!!!!" has been cried over and over and over, and the villagers just ain't coming any more. Nobody believes a liar, even he's speaking the truth.

The weary public, unable to distinguish between hoaxes, facts, stories, fantasies and paranoia doesn't have the option to say "everything is true". Instead, its default position is that "everything is false". Every alleged fact is assumed to be fiction.

We are approaching that world where nothing is real.

What does this look like visually?

In the introduction, we said that some people think that there is one line that matters, the line of truth and lies. Like this:

We then made the point that beauty and ugliness also matter. The captivating and the commonplace. Joining the two graphs, we said that either the one or the other could dominate in an individual's mind, like this:

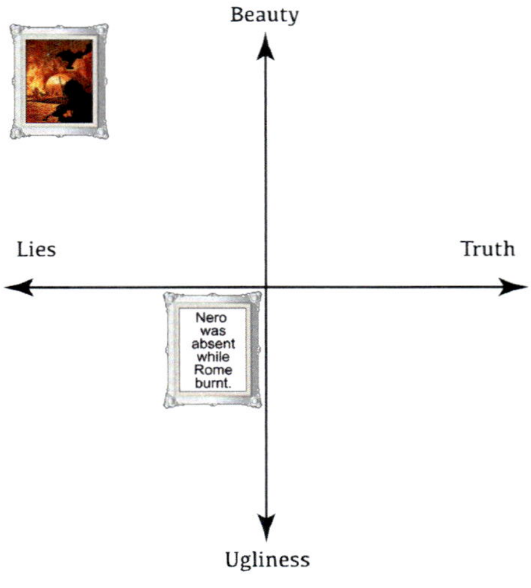

In this world that we're approaching, the world where nothing is real and there is no distinction between truths and lies, lies and truths would be inseparable. But because we're only approaching this world ... and not quite there ... there is still a line called "Lies – Truth". It is just much shorter. Aesop's shepherd boy and Orson Welles and Milli Vanilli and all the others have been shrinking this line, so that for now, "truth" and "lies" are still on opposite poles, but are now much closer to each other than they were before, like this:

This means that we rejoin this graph with the "beauty/ugliness" one, the points we plotted are almost indistinguishable in terms of their truth or falsehood, because the line on which they are plotted has shrunk so much. But the line of beauty and ugliness has remained the same:

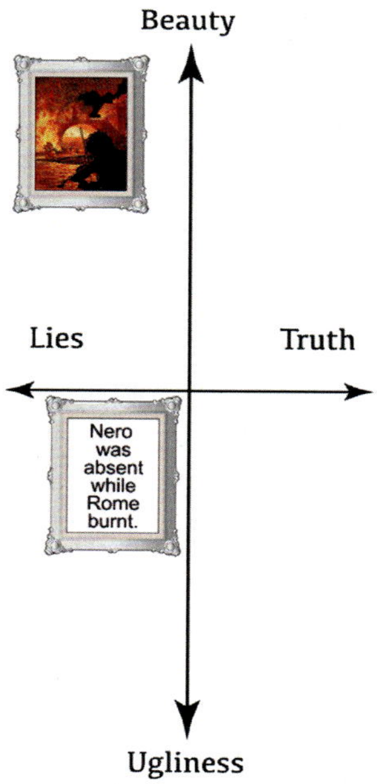

On this combined graph, beauty is becoming so much more important than lies, or truth. Nero with the violin is in the ascendancy, in more ways than one.

In the world we're creating, beauty is steadily weighing more than truth, and ugliness more than lies. A truth is barely distinguishable from a lie, and it needs to be very compelling to be sold. Mere truth is woefully inadequate.

How much chance for Julia or Juan-Franco to cry: "STORM!!!... STORM...!!!," and to get a reaction from the people on the yacht?

It's become very hard to present facts and be taken seriously. Too much fiction has been passed off as fact.

The road to follow is the contrary one: To present fact as fiction.

But Julia, like Esthia, doesn't know this. Julia thinks of her task as one taking place only on a single line, where there is "truth" at the one end and "falsehood" at the other.

She imagines herself and Juan-Franco to be pulling on the one side of a tug-of-war, and the four other people, led by Sam, pulling the other way. The two people trying to pull the existence of a coming cyclone into truth , and the others pulling it towards falsehood.

Soon she will conclude that subtle truth is not speaking for itself and that she needs to add weight on her side by raising the stakes and telling the truth in a very candid way. Shocking and offending if she has to.

Juan-Franco might have learnt by years of trial and error that there's another way to convince, but for the moment he is Julia's secret, silent ally. Not the one doing the talking. Not even showing his colours. This would have been a good time to take out his guitar, with the wind so quiet. But for the moment he's not thinking of that.

5

Hidden agendas

Sam lies on his back with his face to the sun, his eyes half closed behind the sunglasses. Ever since Julia emerged with her crazy account of some cyclone, his mind's been racing, but his stillness belies it.

He just has to be on the island before noon on the last day of the month. Three days from now. That's when the option will lapse. If he hasn't yet signed the main contract, that is. Just arrive, go to the lawyers' offices, sign, and all will be well. A six-star hotel resort snapped up for peanuts from a distressed seller. He could raise collateral against it and use the liquidity to pay the latest margin calls that he'll be getting from his options broker in New York any day now.

He plays back the months and years since the meltdown started. It's all the Fed. Everything was going to be okay, if they'd just done their job. But the Fed caved in to warm feelings of justice and equitability and created the biggest *űber*-bear in three generations.

Sam realises that he's clenching his jaw and that someone may be watching. Lorenzo, the good natured wimp, or the floozy-flaky Erin. They have to be kept calm, in line. He relaxes his neck and jaw, marginally adjusts his head on the towel to allow the tension to escape.

He is very tanned all over, the regular and even tan of sunbeds. He's wearing only a speedo. His body is one of self-reassurance and vanity, with his torso showing the quick wins of a regular gym bunny; large biceps and pecs. It tries desperately to draw the attention from the skinny legs of one who only started exercising when these muscles already needed to have been formed.

It has to turn. It has to bottom out. Not Julia's supposed storm, the real storm that he finds himself in. These people don't understand – Julia and her friends. What's at stake. They've never written eight-figure cheques.

Julia thinks the plane and the condos and the yacht trips just fall from the sky. Her daughter's private school. And she pretends she doesn't care about money, doesn't understand it, only cares about the poor. The poor and their food and their supposed food security. Food diversity and all that endless bullshit. If she understood the market, she'd have known how to help the poor – don't become one of them. And that if food diversity was important it would have had a price and it wouldn't have been necessary to raise awareness or build capacity or fund research because billions would have been pumped into it annually by the multi-nationals.

But he indulges her. Hires yachts to exotic island locations when he could have flown instead. He could already have been there and signed the contract. He tells her to bring her friends. He pays for extra fuel so he can be sure of getting to do what he's come for, even if the wind dies down. So that if it blows, they can sail. This is because he is a good person and he understands that it is his duty to spread his good fortune around. But he is old enough to know that it is thankless pursuit. At the end they will all shake his hand as they disembark. They'll thank him, knowing who's paid the charter, but they'll still maintain that well-disguised snobbism of people who work for non-profit entities, not soiling their hands by touching money.

Well, some people do. And some talk. Good deeds unseen are good deeds nonetheless. But he has to keep them in line and on course for the island or the option will lapse, the margin calls will come with no money to pay, the positions will be closed out miles underwater and none of them will ever sail in these parts again. They may not understand it, but it's in everyone's interest to get this done and get it done in time. They won't understand, can't even if they try. Perhaps Juan-Franco, he was in the game once – but he's a dabbler now. They all live in their insular little worlds of perfect morality without trade-offs or hard choices. They sleep well at night.

Juan-Franco returns to Sam's thoughts. He has that irritating aura of calm and confidence. Sam pictures him talking seriously to the other people on the yacht, and sees them listening. Stopping what they're doing. That confident speaker isn't Erin, nor Lorenzo, nor Angela, nor even Julia. But he has to keep an eye on Juan-Franco. So far he hasn't said or done anything, but he went inside with Julia just after she first started losing it.

Maybe he was checking the radio. Seeing if he could get reception. But Juan-Franco will have to be watched. He mustn't be left alone with the others.

But then, there's always the fact that he is paying. He, Sam. That counts for a lot. They're all here at his cost. So he could pull rank if he really has to.

But it mustn't come to that. He has to keep them calm. Let them know Julia is wrong. Anyone can see it, there's hardly a breath and she's talking about a hundred mile an hour wind. She has to be wrong. Anyone with eyes can see it. But he has to monitor the mood, make sure there aren't conversations between more than two where he isn't present. And then pull rank if he has to. And beyond that? Well there are more things he could do. But it mustn't come to that.

Sam stretches, pretends to wake from a semi-slumber, sits up.

"Jeez, not a breath," he mutters, looking around, "we'll be here forever. That's sailing for you."

He stretches again as he gets to his feet. He sees Erin halfway with her drink and somewhat less than that with her book. Angela seems to be asleep, head on her husband's lap. Juan-Franco is looking out over the ocean, perhaps for better days. For a life.

Sam goes inside and towards the drinks. Julia is sitting, doing nothing.

"So?" he asks.

"Just having a breather."

"You okay?"

"Yea, fine. Wish this radio would receive something more useful than static."

"It'll come back," says Sam, reassuringly. "Anyway, it's so quiet, we can't go anywhere even if we want to."

He sees Julia about to speak, about to engage him, about to call on the inside track that is matrimony – to force a serious discussion. "Mochito's going," he says, turning his body away from her as he starts to fix the drink, "you want some?"

There's a slight silence, and he senses she hasn't given up yet. She wants to talk.

"Isn't this Erin just the dumbest floozy we've ever had on a sailing trip?" he says, softly and conspiratorially, looking over his shoulder. He sees Julia slump with disappointment at his words, at the overwhelming prejudice and snobbism facing her, and shoves a drink into her hand before she has time to recover her sense of confidence in him, or in her own ability to cross the divide.

"Sorry", he whispers as he starts to leave. "I know she's your friend. Just not the sharpest knife in the drawer."

With a smile he is out onto the deck. He settles back in where he was, in the middle of everything and everyone.

"What a day!" he sighs with pleasure as he starts on the drink. "Wow."

He pauses, then continues as if still talking to nobody in particular: "Everything is going to be fine... everything is going to be just perfectly fine."

PART 2

Ugly Truths

6

A NEW SERIAL-KILLER ON THE LOOSE / A SCIENTIFIC REPORT ON PECCA'S

There's the story of the yacht that started with you and me at the radio. Needing to figure out how to convince others to take action, so that we can survive. Six of us in the middle of the ocean. Needing to change course, but not in agreement. We're still there, still trying to work it out.

Then there's the story of our blue planet, a yacht in the sea of space. Millions of species, billions and billions of organisms. A coming storm that science, standing at the radio, knows about. An equally pressing need to change course – but we're not in agreement. How do people perceive and understand? How does science communicate? How can science convince, so that we survive? That story too, is being told.

In that story, it's likewise you and me. Reader and writer. The one writing while the other reads. Trying together to find a form that can convey the message in a way that it can be believed.

Ever since the Enlightenment, for more than 200 years, the truth of science has been ugly. Early on, the science itself was often beautiful; the things it could do, the sense of endless possibility it created. But as machines and pollution began to dominate our world, the beauty of science waned and was only seen on rare occasions. When the Wright brothers first flew. When a man first walked on the moon.

But the communication of its truths was invariably ugly. Ugly and uninspiring by design. Which is okay if you're only writing for people who prefer to read ugly truth. But on our yacht of six, such people can never constitute a strong majority. Just the presence of Sam assures that. And

Juan-Franco doesn't like that sort of thing either. In a democracy of millions of people, it's even more certain that the majority will never like scientific documents. And to change course as a civilization, the strong majority have to listen and agree.

A science-based book about a threat to our world could choose to only start here. In our part two, titled *Ugly Truths* it could choose a title that clearly identifies what we're talking about; maybe over-population, maybe uncontrolled biological experiments. It could say, "Did you know that ...", and then recite the data. Shocking but extremely trustworthy and credible data showing how we're well on our way to doom. The book could contain an undertone of "It's time people are confronted with the truth and stop living in dreamland."

Sure enough, the credibility of the data would show that the content is probably true. In between the facts, rhetoric could be woven in to lead to the conclusion of what the solution should be. But such a book would usually be read by people already agreeing with its premise and not read by anyone else.

So monitor yourself, as we jointly try to discover the form or genre we should be using. How the possibilities make you feel. Two options on how to write a book.

In the one book you quickly interrupt the story of the people on the yacht and say, "I get it, it's a metaphor. Now tell me what it means, what the book is really about." We dispense with the story of the yacht. We never find out how it ends. Sam and Juan-Franco's struggle to bring the other three people around is never resolved. Nero is no longer important to us and we need not find out how the trial of Socrates plays out. Julia doesn't need to go travelling in time and space and we can skip reference to Gandhi and Bruce Willis and *Captain Corelli's Mandolin*. The book starts around here, and I say: "The book is about a new kind of threat that humankind is facing. It's called a PECCA, a Policy-related Environomic Complexity requiring Collective Action.

I explain what I mean by policy related[27] and environomic,[28] then what I mean by complexity[29] and collective action.[30] Then I give two big international examples of PECCA's, say climate change[31] [32] and biodiversity loss.[33] [34] You nod at the former – it fits particularly well with the metaphor of six people on a yacht and an approaching storm. I mention a handful of others such as over-population/over-consumption; acid mine water; global top-soil loss; the sub-Saharan fight against malaria (spraying with DDT); fresh water pollution and depletion and over-fishing.

Tracing the work of decades of experts with reference to exhaustive data, I show that the international community is failing to address these PECCA's, that indeed they're only getting worse[35][36]. Just to tie up the argument, I test to see whether we can solve problems that are not PECCAs but have all the other sub-components. [37] This gives me a chance to create a host of new acronyms like ECCA's;[38] PEC's;[39] PCCA's;[40] PECA's.[41] If we're a little lucky, in a *coup de grâce,* I succeed in pulling off the argument while simultaneously avoiding confirmation bias and managing more or less to stick to the scientific method. In the final chapters I stress how guilty we should all be feeling as humankind and suggest a few corrective actions. But the big message is "Get serious!" The book, full of ugly truth, is read by people who already agree with me – and by very few others.

In the other book we're searching for beautiful truth. Truth it has to be, for science and truth should stick together. But this time the beautiful version. We start with our reader, and I ask, "What do you want? What are you like? What do you find captivating? What can you swallow and what not? How does it have to be presented so you embrace it and love it, even if you didn't previously agree with me – or even if you've never really thought about the topic?" In answering these questions about readers, science is again our ally, for it has studied humans and human perception. It knows what we can embrace and what truth it is that we ostracize and/or (to coin a term) 'ostrich-cize' (to put our heads in the sand, to ensure our emotional survival).

In this second, captivating book, we're searching for beautiful truth. Nero matters and Socrates matters, because they were respectively such exquisite exponents of one-dimensional thinking and because if we put them together, they give us not just beauty or truth – but beautiful truth. Likewise the yacht matters a lot, because it's a metaphor around which a story is built, and what we find in the micro world of the yacht, we find in the macro world of our blue planet. In this second book, we are playful and tell stories. Julia goes travelling through time and space and meets interesting people. We look in every now and again on Nero and Socrates, and meet some other Greek characters like Sysiphus, Zeus and Poseidon. Juan-Franco takes Julia swimming in space, and we watch the earth so far removed that it's only the size of a beach ball. We create an entirely new, custom-made planet to test our thoughts on collaboration and competition, and revisit our best friends from movies and books. We even pay a visit to Disney and *Blackbeard's Ghost.*

All along, just as in the first book, we are interested in the truth and we want to learn. But in this second book, there is canvas and painting, and we make truth the canvas, and beauty the painting. Without canvas there can be no painting.

In this second book there are four parts, of which the second is titled *Ugly truth* – and will have to be written so as to be the canvas. What is the book about in the broad sense? It too is about a new kind of threat that humankind is facing. But it is written in the belief that the most important common denominator of environmental, modern day challenges is that they require convincing communication and interaction with the general public so that collective action can be commenced. Just like on the yacht. For this reason, in the second book, the PECCA's we face become symptoms or case studies of a deeper-lying problem of the communication of science. Therefore, the second book is not primarily about climate change or biodiversity loss or any of the other PECCA's, but about the art of conviction in a divided world. The endnotes however contain a shadow book about climate change and biodiversity loss. The facts, the detail, the pleasing connections between apparently disparate things. But the body of the second book, being about the art of conviction, concedes that scientific reports about PECCA's can't compete for attention with reality TV. For this reason it commences its Part 2, titled *Ugly truths*, like this...

***

It's a busy crime intelligence office. They're always on the run. Crime never sleeps. Sometimes, it's just a stolen handbag, snatched in a busy subway, or a fight over a mistress that escalates until someone is left lifeless in a corner, blood all over the place.

These incidents have always happened, are still happening, will always happen. They don't necessarily make it into the office of half-drunk mugs of coffee and half-eaten plates of sandwiches. Beards not shaven for two days, mascara applied the previous morning. Semi-partitioned desks filled with files and computers and mobile phone chargers.

Only when something happens that is bog or new, or particularly worrying does it come through this door. Police officers throughout the country contend with the crime inherent in life, like coroners tend to the death inherent in life. But when crime steps out of its place and tries to colonize terrain that does not rightfully belong to it, then it comes here to be analysed, studied, dissected, profiled, fought. On this door there is a sign with large letters, saying "VIPs only".

But how do the crime intelligence officers know who the VIPs are? They're not the cops on the streets. If a white rose was left on the face of the motionless man everyone thought had been beaten to death because of a mistress who kept two lovers, does that elevate the crime to VIP status?

The crime intelligence office teaches the officers on the street to look for anything unusual and report it. Just so they can keep an eye on it, and understand if an apparently normal and isolated crime is part of a larger pattern, a network, or a modus operandi. And so the white rose could be reported up the chain of sophistication all the way to the crime intelligence office. But at first, it isn't there. The victim was also wearing two wrist watches, one on each arm. Is that more important than the white rose? Was the rose intended for his lover? Did it identify him to the other lover, who lost his temper? Or did the killer bring it to the crime scene? There is no way to know. The cops on the street, if they're having a good day, mention the white rose in their report, and it goes into the file, and investigation into the case continues.

In the days and weeks that follow, many more handbags are snatched. In subways, streets, shopping malls. Many more women and men keep multiple lovers, and in some cases fights break out between competing lovers who have discovered each other's existence. In a small minority of cases, the fights turn deadly and the cops on the street turn up to investigate a lifeless body covered in blood. None of these bodies display a white rose. The white rose of the first case remains in the file waiting for significance, like a fish hook with bait waiting for a bite.

In the same town, some weeks later, a CIA agent is found strangled in his car, a white rose between his clenched teeth. This case immediately goes to the crime intelligence office – the CIA man was their colleague, and anything concerning the CIA is important. They turn the evidence upside down and inside out, but can't crack the case. No physical evidence of the killer, and no way to know about the significance of the white rose. They don't know yet about the dead lover with the white rose on his cheek. And the police on the streets investigating that case are looking for the second lover, whose identity is known, but whereabouts not. They do not know about the case of the CIA officer.

A few weeks pass. The seasons change. In the north of the country, the first snow glides down, unobtrusive as a whisper. Esthia takes out her waterproof coat and wears it on her walks through the park. In and around the crime intelligence office, it is also colder. The plates of sandwiches are still only half-eaten before people rush to a new crisis, but the mugs and styrofoams of coffee are no longer half full. They are empty.

One morning, a politician is discovered poisoned in his bed, a white rose between his clenched teeth. This information too, immediately goes to the crime intelligence office two recent bodies, two unsolved murders, two white roses clenched between the teeth. Perhaps there is a pattern. There has to be a pattern.

The VIP comes through the door of the crime intelligence office. His status is determined not so much by the importance of the politician, nor by the CIA agent, but by the link between the two. Crime is stepping out of its accustomed place. A case officer is assigned, and she uses state-of-the-art software and databases to start a search. She types in "white rose" and waits while the search is carried out using vast computing power and an astonishing array of databases. And sure enough, within seconds, the first case file pops up. Dead body in a bar, white rose on the face. Eight weeks ago. She feels her pulse quicken and picks up the phone to call the investigating cop on the street, so she can interview him. There's a connection, someone has killed at least three people within weeks of one another. There's a new serial killer on the loose.

The White Rose Killer.

Of course, it wasn't always like this. In times past, there was no crime intelligence office. In the Wild West, in countless cowboy movies, there was only a lone sheriff upholding the rule of law in town. If he was very lucky a *Wanted, dead or alive* sketch might have come from another town to warn him, to inform him of what a known troublemaker looks like. And before that, in Middle Age Europe, there would have been nothing at all. A serial killer escaping from his village could move to another part of the country, another village, and murder again.

Central intelligence about crime grows with time and the sophistication of our world.

But what about pervasive menaces that we don't classify as crime? Things that would prevent our crops from growing and cause famines, perhaps cause new and fatal diseases? Even make our country uninhabitable? How much do we know about them? Are there the equivalent of cops on the street that monitor these pervasive menaces? Is there a central office where the equivalent of crime intelligence can be gathered, analysed, studied, dissected? Are there offices where these dangers can be profiled and fought?

No, not often.

SoBut for a moment, let's imagine them as crimes and use the considerable capacity of the crime intelligence office to profile another serial killer responsible for all the above, and more.

***

In a book aiming not at ugly truth but at beautiful truth, we don't get straight to it and point at the PECCA. No, we start with a story of a crime intelligence office. First, how they would catch the White Rose Killer? And then, how they might use the same insight, intelligence and deductive reasoning to determine the common denominators of the new kind of threat to our planet. In this way too, we will get to the point, but will be involved as people and have a lot more fun.

If we're lucky, the second book is read by people who already agree but also by people who've not thought about this before, or who haven't previously agreed with this line of the thinking.

This is the simple choice faced by you and me, reader and writer. The first book or the second? Ugly truth or beautiful truth?

Because you've known from the introduction where this is going and you're still here, I sense your tacit acceptance of the second book, your insistence that we aim for beautiful truth. I sense that you like Juan-Franco, Julia and Esthia and that you're hoping for a good outcome for the yacht and the planet. I even hear you ask: "What about

a music concert, can we include one? Juan-Franco must sing! And most of all, can we have a happy ending where the yacht survives?"

And I smile and start with the easy question first. "Of course Juan-Franco can sing...how about once in the middle of the book, and reprise at the end? As for the happy ending, well I want that too, but I don't know yet if it's possible."

7

BUT FOR SWELL, A PERFECT DAY

Julia realises that she's been staring at her own shoes for an indeterminate time. Red and white sneakers with one frayed lace and a few loose threads on the one big toe where it's starting to wear. She shakes her head like a boxer on the canvas, trying to beat the count. A few hours ago, Sam was a husband with somewhat divergent interests to her own. Good guy. Good to her daughter. And she was sure that their differences were complementary; they could each go to work and do something completely different to what they did at home. Then come back and be a family again. She'd look at the grey at his temples and think 'maturity', 'stability', and 'man, not boy'.

She looks out at him, settling back on his towel. Moments ago, he was saying something about how everything would be alright, and she felt like running out and assaulting him. Hitting him until she stopped despising him and she could look at him again like she did all the other days that they were together. Kicking him until he snaps back to the person she knows.

Her hands go to her head but stay there only for a moment. She then pictures herself as she'd look, should someone come in now. Somewhat panicky, she tries to shake it. Maybe this is who Sam is. Maybe all the others days were the real Sam, but if the yacht runs onto the reef it will become pretty irrelevant. She's got to keep it together or they'll never listen to her.

Just stay calm, get the facts in order, then tell them. Calmly but honestly. Point out all the risks and why this is vitally important and why they all have to start preparing now. Making a sea anchor, and deciding who does what if it really gets nasty, and who is best suited to what position and what task. Deciding which route would take them away from the reefs the quickest and would leave most space for the cyclone to blow them back.

She rehearses the facts as she knows them, puts herself in the bodies of sceptical listeners, tries to imagine where they might believe her or where they might question her. Listening to herself as she'll convey it all, she thinks she'd believe it if she were them. Sure, the signs are early and not utterly conclusive yet, but it's all consistent with an approaching cyclone. Better safe than sorry, we all know that. It only takes a few hours of preparation. A short while of no sun-bathing, reading, or sleeping. If the cyclone comes and they're prepared, they'll move away from the reefs while the wind speed and swell are still manageable. If possible, they'll sail so the barometer stays steady or rises, meaning they are moving away from the eye. But they'll get as far into open sea as they can, and when it hits and they lose control, they'll throw out the sea anchor to slow the yacht down and start crisis management.

If it doesn't come, they can sun-bathe tomorrow. Drink Mochitos, read books. Juan-Franco can take out his guitar and play his songs, his left hand adept and quick over the frets, his unsteady right strumming simple rhythms. And they can all sing together if they want to, or just listen to Juan-Franco. And ask for their favourite songs.

It's a no-brainer, a slam dunk. They have to sacrifice a few hours of leisure and prepare. It's the reasonable thing to do, and they're all reasonable people. Even Sam. Especially Sam. All his life is about finance and risk and risk allocation and financial instruments that put-it-where-it-belongs. He's just having a bad day.

Julia plays it through once more in a final dress rehearsal, making sure she mentions everything that's certain, uncertain, and likely. She sees attentive faces, occasional nodding, a wink of encouragement from Juan-Franco. Emboldened, she takes a deep breath and walks out.

Immediately she realises she's made a mistake. She was rehearsing what she'd say once she was speaking. Once she had claimed their attention – but she hadn't planned how she would jolt them into listening. Angela's sleeping, Erin's reading, Sam's invisible behind his sunglasses. What should she do? Go back inside and look for a soap box to stand on and clear her voice?

She stops, helpless as a beached whale.

"Hi, Julia, have you managed to get reception yet?" asks Juan Franco, who's turned around to focus the thousand-mile stare on her again, the stare he'd given her moments ago. Now he's spoken louder than was really necessary and caused a stir on what seconds earlier, was a quiet deck.

Before she can answer, a shadow passes over. Julia and Juan-Franco look up to see a single albatross, heading for deeper waters. She realises her moment is passing, again.

"It hasn't started working yet," she starts, trying to speak loudly too, trying to rouse them. "But I'm still checking every now and again for reception..."

She gets stuck, again. She hasn't prepared this part. If it was a music concert the opening band would have been playing now and the MC would be telling them about what an unbelievably wonderful headline act was going to follow. Who's still to come.

"What did you mean just now about the cyclone?" Juan Franco asks. "Should we be concerned? How do we know it's really coming?"

She's forced him into a somewhat transparent talk-show host role, trying to give her an opportunity. Perhaps the others will miss it.

"Well, it has a certain pattern," Julia starts. "Of course no cyclone is exactly like another and one has to leave room for unusual patterns that would be outliers. But in the beginning there aren't any signs of a storm. The wind might be very quiet and the barometer is steady. There may be some cumulus clouds. The swells are likely to be high and getting higher. The frequency of swell is less than ten seconds and gradually increases."

She looks at them, one after the other, tries to remember to make eye contact – as she learnt in a public speaking course in the tenth grade. They're all looking at her now, Sam over his sunglasses.

"Then the barometer starts to drop..." she continues, trying to gesticulate, trying not to be so wooden and frozen, "... the wind picks up and the swell continues to increase in size and frequency. Cloud cover appears. The barometer falls at about one millibar per hour, moving up to two, then back to about a half in the final hours. By then it's raining and the wind is really strong, the waves high and frequent. And then it develops into a full blown cyclone."

They don't respond. The information hovers, looking for its pigeon hole. Where will they put it?

"That's really interesting and useful," says Sam. "Thanks a lot Julia. So you say initially it will be just like a perfect day with some swell?"

"Yes," she starts, so hopeful.

"And how does an actual perfect day with swell look any different?" Sam asks, not a hint of sarcasm in his voice.

"Well it looks the same," she begins. "Unless you look carefully. Monitor the swell frequency and height."

"So it looks the same as a perfect day spoilt by some swell," he concludes.

"Thank you for that. Let's monitor those things and we can take precautions if they appear to be required."

He slips the sunglasses back over his eyes and lies back on the deck. He knows it's a gamble, one or two of the others may yet be undecided, but he sees them settle back into what they were doing before – and knows that he's won another round.

Julia looks at Juan-Franco and sees him smiling encouragingly. It doesn't help the sinking feeling. She notices the word she's chosen to describe her mind-set; 'sinking'.

Suddenly, she feels like laughing. In the way one laughs in a cowboy movie, just before spitting at someone.

Hollow and bitter.

8

What the People on the Yacht Want to Hear

If Julia says, perplexed, "There's a cyclone coming," and Sam says, reassuringly, "Everything is going to be just fine", are we completely neutral as to who may be right? Do we dispassionately continue to gather more evidence and look at the cold, hard facts – in cold blood and with hardened logic?

Some of us manage this regularly – if we're trained as army combat generals, trauma surgeons, sports coaches. Some of us have the ability to be completely rational and clear-minded no matter what happens around us, and no matter what may be at stake.

But on a yacht of six, one such a person may not be enough. There is Lorenzo, the affable banker. Easy-going but dependable. Fun-loving but principled. And Angela, somewhat the same. Erin, slightly in her own world. Fiercely loyal and committed once she's sure about the right and wrong of things. But a little remote and cut off when on holiday. Purposefully avoiding any thoughts of right and wrong, so she can rest.

There is Sam.

There is Juan-Franco, the enigmatic, former business executive who is always touching and rubbing things with his good hand, the one with feeling. The metal of the yacht's rail. The slightly frayed rope. The nylon of the spinnaker at the edge of the snuffer. The ice at the bottom of a cocktail glass.

And then there's Julia, sometimes studiously researching stuff in front of a laptop or from a file with printed matter. Other times steadily adding gloss to her cheeks as she talks and gestures passionately about the poor and the current world order and where the future is going if things don't change.

On this yacht, it doesn't help if one or even two exceptional individuals can see the wood for the trees; it is the majority who must save the yacht, and thus the majority who must receive and interpret information exactly in the way it's been broadcast.

Angela, while listening to Julia and Sam, thinks of her two sons, being looked after for the moment by her mother, until their parents return to take them home. She and Lorenzo will branch off the main road at the white picket fence owned by the neighbour and turn into the drive of her mother's smallholding soon after. Jon and Joshua will have been keeping a look-out through the windows, knowing that it is the day and the hour and that the joyful reunion will likely be accompanied by sweets and other presents. Thus, long before the car reaches the homestead, they will come running out the front door with her mother's Labrador, Goops, in hot pursuit. The thought makes her smile where she lies with a blouse sheltering her face from the sun. If Sam is correct, this is what will happen. Everything will be alright.

If Julia is right, there are other possible outcomes that range from being slightly worse to being utterly calamitous. In the best of them, Angela's mother gets a phone call from her daughter, telling her they're only arriving two days later due to bad weather and that the boys will have to swallow their disappointment. In the worst of the scenarios, her mother's pastor drives past the picket fence, slowly, inexorably and unannounced. He stops in front of the homestead with a dog barking but no kids running out. He brings a message that says the happy reunion has been permanently cancelled and the boys are probably staying where they are for years to come.

Angela cannot be impartial to these possibilities, cannot coldly process the various outcomes as if they all boil down to zeros and ones. She desperately wants Sam's outcome to be true. Thus, if there is doubt both ways, she will tend to believe Sam – even though every moment she lies still in the sun with the blouse over her face, not taking action to change the course of the yacht, is decreasing her chances of survival if Sam is to be wrong and Julia to be right.

The differences in the outcomes are extreme. For Angela's family, everything is at stake. She's totally invested in the outcome and thus very partial. And yet, the research tells us that even when we're just reading a book, we have strong preferences. In a poll run on World Book Day in 2006, the following emerged:

> A truth which has the downside of keeping many true artists poor in garrets and many false ones rich in mansions was universally acknowledged yesterday. It is that most of us crave overwhelmingly a happy ending to a novel; and that Jane Austen's *Pride and Prejudice*

> – in which Elizabeth and Mr Darcy ride to Pemberley in the sunset and live happily ever after – is our runaway favourite of a perfect ending. This truth was confirmed by a poll of public taste to mark today's World Book Day. Nearly 27% cited the ending of *Pride and Prejudice*...Conversely, we long for the President of the Immortals to stop mucking about with Thomas Hardy's *Tess of the D'Urbervilles* and allow Angel Clare to rescue her from the scaffold. We want Cathy to marry Heathcliffe in *Wuthering Heights*. We wish Rhett Butler would give a damn in *Gone with the Wind* and return to Scarlett O'Hara. And we would so like Winston Smith in George Orwell's *1984* to do a Lech Walesa or a Nelson Mandela, overthrow Big Brother and settle down with Julia. These four novels came high among unhappy endings which readers would like to change. So did Boris Pasternak's *Dr Zhivago*, Michael Ondaatje's *The English Patient* and Tolstoy's *Anna Karenina*...forty-one percent are overwhelmingly in favour of books with a happy ending, as against 2.2% who like it sad.[42]

The same is said to be true in the movies.[43] The question is of course very important, because it immediately implies that human decision-making may be much, much further from the zeros and ones of a computer than we thought. Our liking for happy endings is underlined by a visit to a typical Disney plot *Blackbeard's Ghost* and confirmed by some notable examples from literature and the cinema (*Captain Corelli's Mandolin* and *Atonement*). Nero's end was sad for him yet joyous for the people he oppressed. It shows that the beautiful lie ultimately tends to succumb to reality, but at great cost. In counterpoint, Socrates's end seemed sad for him and joyous for his society, but if we look carefully we see that it was actually the other way around.

***

Blackbeard's Ghost[44] was released by Disney in 1968 and the story with its hilarious fun and happy ending was so popular that it was shown for three decades. The same storyline was used in India to make a film called *Chamatkar*.

The hero Steve Walker (played by Dean Jones) is a single, affable man who comes to a small town called Godolphin as the new track and field coach of the local college. The team is utterly hopeless and populated by likeable, but very unathletic geeks. On his first night in town, Walker takes a room in an ancient seaside inn called Blackbeard's Inn. The inn was built centuries before by a famous pirate Edward Teach, also known as Blackbeard, and is now owned by a few elderly ladies known collectively as the Daughters of the Buccaneers. They are in serious financial trouble and unable to pay the bank what they owe on their mortgage. The inn is situated close to shore but on an island, and nobody is sure under what legal system it falls. For this reason, it is coveted by the local crime boss Silky Seymour, who has ambitions to take over the inn and build a casino where the long arm of the law cannot touch him and his gambling establishment.

On that first night, the ladies are auctioning various antiques from the inn in a last-gasp effort to raise sufficient funds to keep the wolf (Seymour) from the door. Seymour's goons are there to intimidate the locals not to take part. Not to bid, so prices stay low. Walker refuses to be intimidated and buys a bed-warmer that had belonged to Anretha Teach, tenth wife of Blackbeard, who died at the stake after Blackbeard denounced her as a witch. As Walker settles into the room, one of the daughters tells him that as Anretha was dying, she'd put a spell on Blackbeard, shouting:

> Edward Teach, sometimes known as Captain Blackbeard,
> When you come to die
> May your body and soul
> Be racked between this world and the next
> Always to be alone
>
> May this curse hold fast and true
> May you dwell for ever more in limbo
> Until such time as there be found in you
> Most wicked of all villains
> Some spark of human goodness!

Walker pays little attention to all this – he's tired and wants to sleep. As he sits down to take off his shoes, he breaks the bed-warmer that he'd put on the bed, and finds in the handle Anretha's book of spells. One is a spell 'to bring to the eyes and ears one who is bound in limbo'. Intrigued, he reads the spell:

> "*KREE KRU VERGO GEBBA KATTO KREE*!"

At that moment Captain Blackbeard (played by the incomparable Peter Ustinov) appears in Walker's room. The spell has brought him back out of limbo. He now has a chance to permanently escape it by doing something good (helping the Daughters of the Buccaneers to retain ownership of the inn).

Only Walker can see Blackbeard with his wild beard, pirate's hat, ever-present sword and ubiquitous bottle of rum. Only Walker can hear him. To all other people he is invisible and inaudible. Only at the end can everyone see and hear Blackbeard.

From there the story develops. Blackbeard uses his invisibility to take on Silky Seymour. In one scene he keeps moving the ball on the roulette table at Seymour's casino, so money can be raised for the inn. In another he interferes with an athletics meeting by tripping opponents to Walker's pathetic selection of misfits so Walker's team can win.

Throughout, Blackbeard is absolutely instrumental in everything that happens – all the way to the happy ending when the inn is saved. Seymour is duly ruined, Blackbeard is released from the spell and Walker drives off into the sunset with the beautiful leading lady, Professor Jo Anne Baker. From the roulette wheel to the athletics meeting to the black eyes that Seymour's thugs get in a fist fight with an invisible opponent, everyone but Walker thinks things are evolving naturally – just happening that way.

Initially, only Walker knows that an invisible force is at work, dramatically orchestrating each and every outcome. At the end it is clear to all.

This is the kind of story we love so much that we watch it for decades. Older generations and younger ones. A story with bad guys and good guys. A previously bad pirate grudgingly becoming somewhat good. The good guys winning and the leading man driving off into the sunset with the leading lady.

We put ourselves in the shoes of Walker, hoping to win over the lovely lady, and in the shoes of the Daughters of the Buccaneers, trusting that somehow, from somewhere, the money to save their home will come ... and we become very partial to the outcome. We see Blackbeard with all his bluster and a tiny bit of vulnerability and know that if we'd been him, we'd want to escape from the curse. We know, even before the story is halfway, which boxes the ending will have to tick to satisfy us. And when it happens in a way plausible enough to keep our attention, we smile and feel happy and sense that everything is alright with the world.

Blackbeard's Ghost illustrates our partiality to happy endings and tells us something about how we like our stories. As will be evident later, it also provides us with the master metaphor for understanding how we deal with facts. But before we get to facts, we visit more fiction – and endings that are more equivocal than those favoured by Disney.

***

Louis de Berniere's *Captain Corelli's Mandolin* entered the world first as a novel, later as a film. The endings are different because of the way we like our stories to finish.

It's set on a Greek island in World War 2. Pelagia is a beautiful and talented young woman living with her widower father, a wise and loving doctor dealing with the sorrows of life by wearing a thick cloak of cynicism and worldliness. She's initially engaged to a young fisherman from the village, strong and loyal but not a soul mate. Antonio Corelli then arrives as the commander of an Italian occupational force. He is full of life and music, seeks out joy and beauty, but he is also young and frivolous. As his men drive with him, singing opera arias in their open vehicle, it seems misplaced and over the top in the context of war and the humiliation of the Greeks, who have lost their sovereignty.

Despite all this, despite her misgivings, Pelagia allows herself to fall in love with Corelli and allows him to occupy her heart, as he is occupying the island. But often she despises him. He is her soul mate, but a soul mate in the making. Until he has learnt about tragedy, until he has suffered, she only likes him sometimes. And yet, in between, there is magic, like the time he plays on his mandolin the song he has written for her.

Beyond the island the war is moving into its final phase. Political alliances are becoming fluid as Germany and Italy lose ground in the war, and decisions are being made that will profoundly affect their lives. Corelli and his men are betrayed by the Germans, lured into a trap and mown down with machine guns – to the last man. Then left for dead. But at the very last moment one of Corelli's friends, who has promised Pelagia to look out for him, throws himself in front of Corelli and partially screens him from the hail of bullets. Pelagia's former fiancé finds Corelli barely breathing and takes him to her and her father. Corelli, after dicing with death, is saved and smuggled off the island, leaving Pelagia behind. The trauma has changed him, made him more serious, given him weight and depth.

In the book,[45] Corelli then becomes a musician and records the song he wrote for Pelagia. He returns to the island and is led by unfortunate events to mistakenly believe that Pelagia has married. After that, he returns to the island regularly but never contacts her. Many years later he finally returns and tells her he thought she had married. She is furious – they have lost a lifetime together. And yet, in the end they walk, holding each other, towards the hut where they'd secretly met all those years before.

This is not an entirely sad ending. It is somewhat happy, somewhat sad. It ticks some of the boxes of the perfect ending we would have created as we read, but not others. Corelli and Pelagia are together, but they have lost so much.

In the film version,[46] the ending was adapted; only a few years pass. Pelagia receives a parcel with a recording of the tune Captain Corelli wrote for her. The island is struck by an earthquake, but life goes on. And then one day, with the lovers still in their relative youth, Corelli returns and they are joyously reunited.

And this is an entirely more satisfactory outcome for us emotionally. It's the way we wanted it. It is plausible, and while the threat that they will never be together again is there until very near the end of the film, it ultimately all falls into place.

A similar dynamic existed in another great World War 2 story that became a film, Ian McEwan's *Atonement*. The book is written from the perspective of a female writer, Briony, who misreads events around her in her teenage years and causes her sister's lover to be wrongly convicted of a serious crime. In the process she destroys their lives and separates them. Many years later, old and dying and wishing for some atonement, she writes the story of what happened to her sister Cecilia and her sister's lover, Robbie.

Robbie is ultimately released from prison to fight in the war, Cecilia becomes a field nurse. They see each other only briefly before he is off to the front, facing the worst of the Allied meltdown at Dunkirk. He's in a bad way physically as he arrives at the coastal town, waiting to be evacuated. In a breathtaking juxtaposition, Robbie stumbles past the silhouette of a deserted merry-go-round, looking for water – but there is none. None of the laughter that in earlier times must have rung out from the merry-go-round. No life. Not even a drop of water for a desperately thirsty man.

And yet, in time Briony meets up with Robbie and Cecilia again. Both are on leave and living together, starting to put the horrible past behind them. As they part, Briony realises that Robbie was innocent all along, and promises to fix the record.

This appears to be the end of the story of Cecilia and Robbie. Mostly happy, despite all the misfortunes. But as the book winds down, Briony confides that in real life, Robbie died at Dunkirk, while Cecilia drowned when a bomb destroyed the water supply above the Balham underground station. She had only given them a happy ending in the novel because she was the one who had prevented it in real life. She writes about their demise in real life.[47]

How could that constitute an ending? What sense of hope or satisfaction could a reader draw from such an account? Who would want to believe that they never met again, never fulfilled their love? Who would want to believe that, except in the service of the bleakest realism?

We don't like our fiction like that. Robbie and Cecilia, never reunited. We live uncomfortably with Corelli and Pelagia, reunited in old age. And as long it convinces us, we love Corelli and Pelagia, together as semi-young people, with time to erase some fairly short lived horrors with joy and happiness in the decades to come.

We want happy endings.

***

In Rome in AD 64 , the great fire would end happily for Nero if only it would burn long enough to destroy all that is vulgar and usher in all that is wonderful. And yes, it Rome, AD 64

burnt and burnt, spread and spread. Over the narrow passageways to the neighbours, then to their neighbours. From the one district to the next, to the next.

In daytime there was mostly smoke, but every night the red and orange and yellow would gnaw at the foundations of Rome. The smell was strong, mixed, and hard to pin to anything except for the certainty that it emanated also from the incinerating flesh of those who'd failed to get out in time. Roman citizens, visitors, slaves, dogs, rats.

As the sun rose after the first night, everyone close to the flames was either dead or had evacuated. There was a stampede early on because people realised they had to get out immediately; the weaker, the younger, the friendlier and the less protected fell down and often didn't rise again because wave after wave of terrified victims ran over them. But by sunrise, everyone knew and kept a safe distance. Except for the wailing mothers and distraught fathers, the bereaved husbands and wives, the orphaned children braving the proximity of the flames to search in the alleys for someone who miraculously might still be alive.

It was strangely quiet from that second morning. The cries had died down, the people had run, there was only the occasional, razed house collapsing as the flames tired of it and prepared to move on. Sounds from deeper in the city carried and could be heard in the emperor's palace on the Palatine Hill. Mostly dogs barking. But for the most part, the life in the surrounding districts, still burning, observed an eerie silence, as if to respect the dead.

The flames were too overwhelming to fight, and it seemed they would just keep rolling until the entire city had been devoured, all of the country, all of everyone's lives. Two days of helpless misery turned into three, five thousand homeless people to ten, and so on and on and on and on. Three days, four days, five days. By the sixth, ten districts had been destroyed and only a third of the city remained. Three districts had been razed to the ground, the other seven were a jumble of destruction and survival, extinction and conservation.

And then, the fire failed to spread from one neighbour to the next, and to the next. From one district to the next. By the evening of the sixth day, Rome was just smouldering, and people began to return to what used to be their homes, their neighbourhoods, their circle of friends. They scavenged for remaining possessions and looked for the remains of loved ones. The silence was displaced, there was wailing and crying, but some just stared vacantly ahead in mute silence.

People moved away to survive. Then the slaves moved in, in large teams, and started clearing the rubble. It was time to rebuild, but in a different style. One that would not be susceptible to fire. Broad streets, spaces between homes. Only brick buildings. In the middle, a large space of about 200 acres was cleared, and gardens were constructed. It was said that the emperor was planning a new palace complex to be called the *Domus Aurea* – the Golden House.

And then the astounding tales started to go round of what the emerging wonder would look like; gold everywhere; semi-precious stones in the ceilings; frescoes; a fresh water lake; also forests, vineyards and flocks. About 300 rooms, but not a single bedroom. Just entertainment.

In hushed and awe-filled voices, those in the know told about white marble everywhere. Pools and fountains, and a canal which would stretch all the way to a large lake some miles away. The ceilings had mosaics.

Everyone wondered where the money would come from. So many people still had no place to stay. When they heard the provinces were being taxed, there was a collective sigh of relief that it wasn't them. But still – all that money.

There was talk of a giant statue, to be made by an artist from Greece. Zenodorus. And sure enough, soon he could be seen working on a gigantic structure, with the help of dozens of slaves. It was said to be the sun god Sol, and was being made in bronze.

For a very long time they worked, but when it was done, it had the face not of Sol but of Nero. As high as 20 men standing on one another's shoulders, it was at the very end of the Via Appia.

When the new palace was done, Nero looked at it without much emotion and said: "Finally I am being housed like human being."

But the people who'd lost their houses in the fire didn't feel like that at all.

The ending was happy for Nero, sad for them. And very sad for many others, who'd lost more than just their homes.

***

Half a millennium earlier in Athens, a recurring dream was trying to prevent Socrates from creating a lasting tragedy.

"Make and cultivate music," the dream had said, over and over. When he was still a stonecutter, as a young man, he'd first had it.

"Make and cultivate music."

And he recalled having it again when, in his 37th year, he was fighting in the Battle of Potidaea and the siege just went on and on and nobody was sure if they'd make it out alive. The summer there was hot and dry, and most days he'd look out over the olive trees and pale grass and as the sun went down and the stars became clear, he'd think about his life back home. The half behind him and the half in front of him. He'd wonder why he was doing what he was doing. His fellows would be heading off for a huddle somewhere to drink wine and eat around a fire and laugh at obscenities and joke until they staggered back to where they slept. And all the while, Socrates would sit staring at the stars or at the fields, now invisible in all but the most basic outlines.

Years later he'd tell his students that the worthless only live to eat and drink, the worthy eat and drink to live.

Always there was the divide; him with his heavy thoughts like the world on the shoulders of Atlas, they with their light-heartedness, their lack of weight – like a morning flower. And no way to bridge. They felt it, and while they respected him and perhaps feared his judgment and intellect, they didn't befriend him. No discussion was needed, and none was had. When business ended and pleasure began, they left him behind and went off on their own. And the divide pointed fingers both ways, at the huddle and its outsider, exclaiming: "Oh you frivolous excuses for human beings, your lives aren't worth living!" and "Oh you sad and sorry creature, if only you had a life!"

Always they judged him and despised him, always he judged them and despised them. They who always thought they knew everything about everything.

"I am the wisest man alive," he would say in later years, not just to his students but to anyone who would listen, "because I know that I know nothing." And the disdain and misanthropy and elitism would be all over his face as he said so.

But the soldiers in the huddle agreed with him that he knew nothing. Not about anything about anything that mattered; the joy of women, the pleasure of wine, the light of companionship, the forgetfulness of uproarious laughter. All the things essential in a war. Of all that, Socrates knew nothing. And if that meant being wise, they were happy to be ignorant.

Those were the days and nights of Potidaea – much the same. A siege is a waiting game. One day is like the next, and soldiers make their peace with it, for when it would change it might change for the better – or very much for the worse.

Socrates too, was okay with waiting. He was thinking. If he went home, he'd still be thinking. But then randomly, some night, suddenly, it would come to him, the directive from some god to turn to the lyre. Apollo perhaps, or even Dionysus. Apollo because he was the god of music, or Dionysus because he was the god of wine, joy, theatre and peace. Everything Socrates despised on bad days and failed to value on good ones.

Make and cultivate music.

But it was so utterly against everything in him. He hated tragedy – it seemed like a celebration of human folly. Poetry he avoided altogether, and he only went to the theatre if the plays of Euripides were being performed. They were entirely logical and they were about things that mattered.

Pondering the instruction of the recurring dream, was he supposed to take up the lyre and perform the lyrical poetry that made women cry, and look at the artist with a glimmer in the eye? His whole world was built on logic and deep contemplation. Removing himself from the vulgar, not drenching himself in that sea of deception and illusion. And so he disregarded the dream and hoped it wouldn't return.

But it always did return. Back home after that war, and again during the next wars at Delium and then at Amphipolis, where Athens had suffered a rout and Socrates and one comrade were the only ones walking home as all the others ran like petrified, wild animals. Socrates walked home, past many victorious Spartans, and shouted at them that they would die if they came anywhere close, and they turned and chased instead the ones who'd dropped all weapons and were running like rabbits.

When death was close, the dream seemed to return.

In the years after that he married Xanthippe, but that was a big mistake. The illogical moods and inane emotion was suddenly inside his house, and much harder to escape. It was as if the soldiers of the huddle with the frivolity slept in the same bed as him, under the same roof, every night of every year.

He told his students, in later years, that they should marry and hope for a good wife and happiness – or another wife and the life of a philosopher. With Xanthippe, the choice had been made for him. Her dark, restless eyes, the fidgeting fingers, the accusing drop of the shoulders – another testing situation, another disappointment from her husband. Philosophy was the last realm he could escape to. His father died at about that time, and Socrates inherited enough to stop working. That's when he began teaching the youth free of charge, and being a philosopher.

He was in his fifties, and it was too late for a fresh start. A life of opposition and defiance had developed for decades – all he could do was perfect it. Athens wasn't a happy place, things were going wrong. It was an ideal time to be a contrarian.

When his grandfather was born, democracy had finally come to Athens after decades and centuries of war, expansion, serfdom and suppression. There was peace for a while, and prosperity, but then the Persian wars started and everything went the wrong way again. Only when Socrates' father was a small boy did those troubled times end and the real prosperity and progress begin. His father cut stone for the Parthenon and other magnificent buildings. Not important to Socrates, but a source of pride to Athenians.

For fifty years, there wasn't much war, but then Sparta came along and just wouldn't go away, or die, or implode. Thousands and thousands of talents were thrown at the war effort, with little to show for it except decay, creeping poverty and high taxes. Athenians whispered to each other about discontent and about changing things ... that this just wasn't working and that something had to change.

And so, as Socrates came into his prime as professional philosopher and Xanthippe bore him the first of three sons, Athens and its democratic heritage began to wobble. Everyone felt the golden age slipping away, nobody knew how to get it back. In 411 BC, a few years before, there had been a short-lived coup, but democracy soon seemed to fight back and restore itself. And yet, the irresistible force was meeting the immutable object, and something was going to have to give. Democracy was leading only to degeneration and continued war.

In the early morning and at sunset, Socrates would often look up at the Acropolis for which his father Sophroniscus had cut stone all those years before. In those halcyon days, one could look up and measure the health of Athens by the state of the Acropolis – progress was constant as first the walls and then the temples were rebuilt in the magnificent marble, in symmetrical lines. And the crowning glory, the new Parthenon, kept rising for another two decades afterwards. The Acropolis complex seemed to Socrates a symbol of all that was ordered and logical, considered and designed – just the way a society should be. Watching the day break over such organizational harmony seemed to him the best way to start. At sunset the last rays from the horizon illuminated the upper reaches of the Parthenon, bathing them in light, while the lower parts were indistinguishable from afar, as shadows crept inexorably up the columns. At that time of day, it seemed symbolic of everything that was going wrong.

Democracy and the Golden Age had coincided, lived together. Now they were in danger of dying together. To Socrates, what was wrong was still exactly what he experienced in that first campaign at Potidaea: There he was, the wise philosopher who did everything in moderation and stayed to the side. And then there was the huddle of drinking, laughing, joking soldiers next to the fire. The inane, who never thought about anything. What was wrong was that everything had dropped to that level, including political discourse. Crass simplification was the order of the day, and citizens thought they could change rules to suit them. Laws could be made or waived to placate them and their emotionally-fuelled desires.

Constitution? Screw the constitution. And the same to logic and wisdom. The enemy was at the gate, both militarily and economically, and amoral shortcuts were the order of the day. Ill-considered decisions were taken and implemented with far-reaching results, to be followed by regret. What was wrong, thought Socrates, was that there wasn't a Philosopher King.

And the conflicts just rolled in, and Socrates just continued to disagree, like he'd always done and always would do. By 406 BC, he was a senator in the National Assembly, the Boule. It had a rotating chair.

Eight generals had just won a great and unexpected victory against Sparta. At the end of the naval battle, there were many sailors in the water because some Athenian vessels had sunk. And there were other ships and comrades still at the mercy of Sparta. What to do, save the one or the other? The generals used the bulk of their ships to try and relieve their comrades still under threat and left a few vessels behind to save the Athenians in the water. Then a storm came, and many men drowned.

Their bereaved families back home heard half a story of ships sailing away with men in the water. The uproar and backlash were immense, and soon the eight war heroes were eight potential war criminals, and being tried in a criminal court.

The Constitution prescribed due process, but the people were angry as the matter was brought before the Boule. As luck would have it, Socrates was presiding as the chair for the day.

"Enough, enough!" cried an opportunistic politician called Callixeinus. "We've heard enough, let us vote now on their guilt!"

And yet, the trial was still underway. But the people were angry.

"Let us follow the Constitution!" shouted Euryptolemus. "The Constitution says otherwise!"

There were shouts and accusations, and Socrates had his hands full to restore order. "LET ANYONE WHO OPPOSES AN IMMEDIATE VOTE RECEIVE THE SAME PUNISHMENT AS THESE DESPICABLE GENERALS!", shouted another opportunist, implying only one possible outcome to the vote. There was a chorus of angry agreement.

Euryptolemus and his supporters looked nervously around at the baying crowd and realized that continued opposition could be a life or death decision.

"SO LET US VOTE THEN!" shouted Callixeinus, his voice unable to hide his excitement at what was unfolding. Socrates looked at them. The despicable ones being swept away on the wave of the moment. The huddle of frivolous soldiers.

And disagreed.

"I will do nothing that is against the law," he said solemnly. "This meeting is adjourned."

And for a little while the lives of the generals were saved. More importantly, for a little while rationality held sway, the rules were obeyed, and the real enemy (as Socrates saw it) was kept at bay at the gate. Xanthippe would sometimes start shouting and throw household items about, for no apparent reason. She'd claim out of the blue that he didn't love her. The next day, she'd be calm again. She was as variable as the weather. And in the same manner, the Athenians would suddenly hate someone and want to kill them like the generals. But then, soon after, they'd regret their actions and blame someone whom they'd agreed with at the time, like Callixeinus – who after the execution of the generals was completely ostracized to the extent that he died of starvation in front of everyone's eyes.

From his youth Socrates had disagreed with those around him, but when he became a well-known philosopher, his contrarian habits became common knowledge. And by chance, the one day he'd chaired the Boule, he had an opportunity to disagree most spectacularly. As regret set in among his detractors, he was quick to rub their noses in it. "I am the wisest man alive," he'd say to anyone who would listen, and many others too. With the clear and intentional sub-text saying: "You killed them, you bunch of fools, not me. You sorry excuse for human beings. The uncontemplated life is not worth living. You don't deserve to be here, you are a disgrace to Athens and all she stands for."

It was the truth, so Socrates told himself. They needed to hear the truth. The truth, the whole truth and nothing but the ugly, bare truth.

The die had been cast – sometime there would be another event like the one with the generals, and they would want to bend rules again. If he stood in their way, they would come for him. Perhaps they would come for him anyway. Being so clever and aloof and such a gigantic pain in the ass. More than anything, a reminder of what they had done. Like Callixeinus, Socrates was an unhappy reminder of a less than glorious episode. A reminder that they could well have done without.

And so the last seven years of Socrates' life commenced. A very unhappy ending was the chariot in the inside lane. More and more often, the recurring dream that he didn't understand, tormented him. This one failure that would cause his death and leave a ruinous legacy to Athens and all her offspring.

"Make and cultivate music."

What did it mean beyond the obvious? Which of the gods did it come from, and why did he have to do it?

The philosopher would never understand, and only on the brink of his carefully orchestrated, sad ending would he start to obey – in a humble and coy manner.

9

Juan-Franco's sadist muse

Esthia leaves her coat behind as she walks out of the office and back into the street, retracing her steps. She stops at the same coffee vendor and this time buys a cappuccino. From the subway below vapour rises, enveloping her like the heroine in that one movie she once saw with Juan-Franco.

The coffee is too hot to drink, and she walks cup in one hand, the other withdrawn into the sleeve of her jersey. The sun is up now, but the thaw is slow. The chill is already in her bones, has been there since she saw that mass of white cloud on her computer screen. She feels it creeping onto her fingertips and settling on her neck like an unwelcome insect. It'll be all over her soon. Depending on the cyclone, perhaps it will stay.

In the park, she takes the first walkway splitting off from the one she uses every day to go home. She continues to walk, somewhat randomly. She goes past the empty seesaw, then the swings. Two kids and an adult watching. Esthia is at the small pond then and sits down on the empty bench. Its cold penetrates her clothing and spreads to her thighs and lower back.

Pablo Neruda. She wonders where in her home the anthology is. Whether it's even there. Perhaps it was lost in one of the moves from one apartment to another. Raising the cup that's no longer piping hot, she sips. Remembers walking under the rain with Juan-Franco and how drenched they were when they arrived. And Greece and the ferry. Telling him she couldn't do it. Not in the flesh.

She pulls her feet up below her, puts them on the bench, places her arms around her knees. She wants to feel, wants to be cold. Wishes for snowflakes to start falling on her hair and face and shoulders and hands. For the sun to warm her cheeks just that tiny little bit. And yet she's going numb, from her fingers into her wrists, down her neck onto her back, through her thighs and lower back into her being. And from there, where the chill settled when she saw the cyclone, the numbness spreads.

He can't have mobile coverage or he would have called. She needn't check, she knows. And so all she can do is have another virtual discussion with him. Like playing chess with herself.

She's done this many times. Every day? She wonders. Perhaps not every day. But on some days, many times.

It's been so perfect. Everything staying in its place. Seeing him in a coffee shop like the one below her office, then months or years later at an airport lounge. Walks three years apart, one in Hampstead Heath, the other down the steps to Clifton Beach because their schedules happened to coincide.

Once she even managed to attend one of his performances in that small, famous venue in the basement. She saw the hundred or so people, young and old, rich and poor, listen to the poetry he read, absorb the thoughts he shared, delight in the stories he told and the songs he played. Lingered afterwards to talk in private, hoping something of him would rub off and stay with them – as it's stayed with her. She's had what they want – the inside track. The more illuminated view on the general enigma. They all sense there's so much more to Juan-Franco than what can be seen and felt in 90 minutes on stage or in three minutes of private conversation ending with an exchange of business cards. That in the juxtaposition of artist and businessman there has to be vast and rich contradictions. Passion and serenity. Wisdom and naivety. Philanthropy and asceticism. In that song about the businessman and the beggar, he was both. Every day he is both.

In the closing minutes that night, Esthia looked at the faces around her. Especially the women. Saw them frowning from time to time as Juan-Franco's intensity built towards the climax of his performance. As his strumming became faster and more urgent, the content of the songs more profound, as his voice rose in volume, hoarse and full of integrity. Imagined these women wondering who feeds him emotionally. Who the Other is who inspires and energises him, and tortures him to create the rawness in his songs and their rendition. His sadist muse.

Now, thinking of the yacht, she remembers how a blue spotlight fell on Juan-Franco's face as he sang a long note at the end of a song. There was sweat on his brow, stubble on his cheeks, his hair was long and free and framed his tanned face. His eyes were half-closed.

Depending on the yacht and the cyclone, this might be what she remembers, always remembers. That moment, that facial expression – half present, half elsewhere. That light and its shade.

Esthia absent-mindedly looks at her watch, doesn't notice the time, but wonders what her two children are doing. Whether everything's okay.

When someone walks past in leather heels on the paved walkway, she realises that time without thought has elapsed. That between thinking of her children and now, seconds or minutes have slipped by.

At that crescendo, Juan-Franco's show ended. There was applause, and he played one more song as an encore, then stayed and talked to everyone who wanted to. But before someone else could play, that night or the next, the stage would have to be cleared, cleaned, and rearranged for what would follow. This is what just happened to her mind in the minutes or seconds when she wasn't thinking.

She looks east, and the sunrays are so weak and diluted that she doesn't have to squint. The coffee cup is empty – she must have finished it while she was remembering. The cold has conquered her, and she knows she'll start to shiver soon. Rising stiffly, she walks the first few steps back to the office with a hunched back, waiting for her muscles to unfreeze.

Perhaps she will email him. Write him. Perhaps write him but not email him. About the day walking in the rain and about Greece and the day on the ferry. About being the Other.

Depending on the cyclone, he might not get it if she emails him now. But then she would have taken the chance to say something while it was possible and probably feel better.

Or he might actually get it when the cyclone has gone – and everything would be out of place, and the perfection over, and she would be sorry she sent it.

10

Green Sysiphus

Juan-Franco grew up in a city that expanded very quickly at the time, and in the preceding years. Once, people wanted to live near its centre, but as it grew and the old residential buildings from the early days started to decay, young people moved to the peri-urban areas further away. Some chose to live there on smallholdings while others built houses in the new neighbourhoods springing up on the edge of the budding metropolis. Most planted trees when they finished building their houses, contributing to the largest man-made forest existing anywhere. There were a good number of swimming pools in the warm climate. The combination of trees, water and the absence of natural enemies drew a great many birds.

Just a mile beyond the place where Juan-Franco's parents built their house in a quiet cul-de-sac with an unusual concentration of university professors, there was an untouched wetland where he ran round barefoot with his boyhood friends on the footpaths, dug out black clay from the edges of the water and kept a careful lookout for snakes. It was a place of sanctity and wildness that had a central place in his frame of reference.

And yet, it was a city. Millions of people lived in it and needed to get round. Roads were built, and schools expanded. Children were taught to look carefully both ways for oncoming traffic before crossing a road. Huge storm water pipes were laid on the edge of the wetland, and gutters built above them on the side of the roads.

Juan-Franco got a bicycle when he was seven or eight. Once he'd been taught how to avoid the cars, trucks and buses, his parents set him loose. Suddenly his world was much bigger. There was a community library just three miles away, that opened up the sights, sounds and smells of a multitude of worlds beyond the horizon. Soon, the youngster was making his way there twice a week to exchange books. Passenger buses were the greatest hazard on the way. Most were old and poorly maintained. The drivers didn't give cyclists a wide berth and Juan-Franco had to concentrate hard when he heard a bus approaching from behind so as to keep his bicycle close to the kerb and travelling in a straight and predictable line.

One late September afternoon after school and sports, he was on his way to the library and heard a bus approaching. He didn't look around, knowing it would only cause the cycle to drift out of line. An intersection with traffic lights was close ahead, and it became progressively harder to keep the bicycle steady as his speed dropped to observe the red light. But he managed, stopped. The bus pulled in next to him, crossing the white line before stopping so that Juan-Franco was somewhere near the middle of it once they were both stationary. They both waited for the light to change. As it changed, the driver let go of the clutch and stepped on the accelerator. The rings of the bus engine were utterly shot and with the exhaust pipe of the bus being right next to Juan-Franco, somewhere near the middle of the vehicle, a thick cloud of sooty smoke was emitted straight into his face before he could move. The bus drove off blithely, leaving the young Juan-Franco in a fit of coughing, gasping for breath and seething with anger.

Something changed in that moment of incongruity; the bus could make him cough, could make his clothes smell of that smoke, and yet suffer none of the consequences of its actions. The passengers breathed clean air, the driver too. Perched high up on the seats, they looked out well over the head of a boy on a bicycle and more than likely didn't even know about his discomfort. But either way, they were out of there, leaving their adverse consequences behind as they moved to the next intersection, where perhaps the process might be repeated with another helpless cyclist.

Some years later, developers bought the wetland and turned it into a circular shopping centre with an artificial lake in the middle. In the centre of the lake was a stage where rock concerts were held. Juan-Franco was about to move away, but his special place and its former inhabitants were gone. He went to college and did well. His post-graduate studies took him abroad, to meet Esthia. A sophisticated city where would-be ballerinas practise their moves next to the walkways and spirited buskers are found in the public parks.

The problem of the bus still bothered him, but on a global scale. It bothered other people too, most particularly the Intergovernmental Panel on Climate Change. It issued a report saying there was trouble ahead. The follow-up reports every five years or so usually said things were getting worse and that the prognosis was more negative than before. By the time

Juan-Franco started working for a boutique law firm in London, mostly representing small island states in the international negotiations held annually under the auspices of the UN, about five annual negotiation rounds were complete and a pattern had been set.

After a year of preparation, the international community meets for two weeks. Each member country sends a team representing it. They work for months to prepare for the meeting, drawing up a Position Statement. It is drafted carefully and precisely, in a way that is fair and that fits in other policy objectives. Each country looks for allies and lobbies beforehand for an outcome that best serves its interests. Negotiating blocks are established and new acronyms to describe them. There are BRICS, and AOSIS, and G77s, and G8s, and OECDs, and many others.

Non-profit associations caring about grasslands and birds, or grasslands and insects, or desert flowers, break their budgets, appoint experts to ascertain how the impending menace might affect what they love, then appoint policy officers and lobbyists who work hard to win the right to attend the international conference as observers. The budgets are strained further to organise side events where non-profit associations demonstrate the potential effects to the world. They print promotional material, T-shirts, and make DVDs that they hand out free of charge.

At the venue, the host country has been preparing for months and years. Looking at logistics. Wondering about protocol and bringing in the best facilitators. Organising food and drink. Show-casing pilot projects that address this new issue. Making security watertight for the heads of state due to arrive in the second week.

Then the big day arrives. From all over the globe, they travel to the one city. At the conference venue, all is ready, tidy, clean, secure, and facilitated. The media have gone through an accreditation process, and only the legitimate are allowed in. The opening plenary sets the scene, and thousands upon thousands are filled with hope, seeing so many others from every corner of the globe, all gathered for the same purpose.

The first week passes, and the media report very slow progress, with many points of dispute about what is fair and whose responsibility it should be to do what; and about how to divide the global sacrifice. The poor say the rich should do more, while the rich say everyone should do as much. Some non-profit organisations say it doesn't matter if the sacrifice is divided fairly, what must be done must be done and that if some people want to ride on the efforts of others, that is okay.

In the second week, the heads of state arrive. Everyone is filled with renewed hope that these influential Heavy Hitters will sort everything out. Will use their charm and magic to succeed where the assortment of negotiators and facilitators have failed.

By the second last day, it is clear that no agreement will be reached. But everyone has come so far, and so much hype has been created, and so many millions of people are watching, that going home empty-handed is unthinkable. So the Heavy Hitters get together and say that a 'Road Map', a plan to move forward, is required. For the next six months, negotiators continue to prepare an agreement to be circulated within another three months, so that in one year, when the next Conference of Parties gathers, a binding decision can be taken to ensure the menace never impacts on humanity in a way that is 'Beyond Reasonable'. The negotiators are tasked with finding a legal definition for 'Beyond Reasonable' that all can agree upon.

On the last day, the media are called in and the Heavy Hitters from the G7 and G77 and AOSIS and other constellations triumphantly unveil the Road Map. Non-profit organisations wondering how a so-called Road Map will benefit grasslands or insects or birds or desert flowers in the next twelve months express their reservations, but generally there is optimism. In effect: we haven't done it, but we know now how we're going to do it. It's a matter of time. We have a Road Map, a 'Plan'.

A year later, another city hosts the COP, the gathering that so much was expected of a year before. During the year, little progress was made towards agreement on the Road Map. Even consensus on a definition of Beyond Reasonable is proving difficult. Some countries say the time span used to determine what this means should include the epochs preceding human life on earth. If the world was very different in the primeval past, is it Beyond Reasonable to see it returning there? Others say we should confine ourselves to conditions that historically accompanied human life. Negotiations stall.

In the second week, the Heads of State arrive, and there is a renewed sense of hope that they will use their charm and magic to succeed where the negotiators and facilitators have failed. Some of the Heavy Hitters have lost elections and have been replaced by others – perhaps they are the magic ingredient that was missing before.

On the second last day, it becomes clear an agreement will not be reached.

But everyone has come so far, and so much hype has been created, and so many millions of people are watching, that going home empty-handed is unthinkable. So the Heavy Hitters get together and adopt large sections of the Road Map for "Early Implementation" and assure the public that the remainder of the Road Map has made meaningful progress towards a solution. This, they say, will be adopted at the following year's event.

The media are called in and the Heavy Hitters from the G7 and G77 triumphantly announce the progress. Non-profit organisations, AOSIS and other organisations wonder how the alleged progress will benefit grasslands or insects or birds or desert flowers or low lying islands in the next twelve months. They express reservations, but generally there is an air of optimism. It's just a matter of time. We have a Plan, and large parts of it have been agreed upon already. It's merely a matter of time.

And they meet the next year, in another city, in another country. And then another year later. And in subsequent years at the 5th and 8th and 11th and 12th and so forth. Twenty consecutive years they meet, and wonder how the sacrifice should be divided, and what would be a fair way to avoid the 'Impending Menace'.

Over and over they do this, over and over, again. After his third year of personal involvement, Juan-Franco started looking for other solutions. The process was making him think of the myth of Sysiphus, who for his severe misdemeanours was sentenced by the Greek gods to perpetually and fruitlessly roll a boulder up a hill.[48] Roll it nearly to the top, and the boulder rolled down again. Each time, with monotonous regularity, Sisyphus would walk down and start yet over again. Each time he rolled it nearly to the top and the boulder would roll down again.

Sisyphus would walk down and begin all over again. roll it almost to the top. The boulder would roll down again.

Over and over; over and over; and over and over... for all eternity.[49]

In looking for international consensus on behalf of the small island states, Juan-Franco had become Green Sysiphus, rolling an environmental boulder up a hill, only to watch it roll down again. Over and over.

He started looking for other solutions. Maybe financial derivatives. After all, what the financial markets do is to establish linkages. If it snows a lot in Kazakhstan at the wrong time of year, wheat in the US may become more valuable. If a product could be created for people to buy and sell, the markets would establish the linkages between a changing climate and the impact on man. Conviction would become easier. Juan-Franco did a post-graduate certificate course at the London School of Economics.

Along the way, he'd grown very close to Esthia, then drifted away again. Music became a bigger part of his life as he continued to ponder on how to solve the problem of building an international consensus.

He sits on the deck now, sure that Julia is about to emerge. She's going to try and convince the other four again, soon. But every time she fails, it becomes a more difficult task. Not only is she endlessly rolling a boulder up a hill, she's making the hill higher every time. She's going to emerge again, try again. Juan-Franco knows that he has to help her in some way. Perhaps by using one of the approaches he learnt after giving up on Green Sysiphus. There is the way of Gandhi, and the way of Bruce Willis.

Maybe one of those. But more likely a song.

***

Inside, Julia takes up her pen, then pauses again.

A seven-year old with blonde hair and a freckled face interrupts. It's her daughter, Stella. She's thousands of miles away, and she says urgently, like one repeating herself in exasperation: "Mom! Listen to me! If this is true, am I going to see you again?"

Despite her absence, she's been the elephant in the room, the one who causes Julia's throat to dry and her pulse to race. What if? What if? She's urging Julia to be part of the conversation, to go into the equation and be part of the decision-making on what's to be done. Perhaps she's even imploring: "Forget the facts! Just walk out there and talk to them about me! Say everything in your being wants to see me again and wants to watch me graduate one day, and wants to be at my wedding!"

But like Socrates, Julia takes a deep breath and declines. Her credibility as a scientist is on the line.

She looks the writing pad up and down. It's A4 size. Should she try to get everything on one page, or just write and let it be as long as it wants to be? Unsure, she starts, resolving to be concise, as concise as possible.

She has to tell them exactly what's about to happen. Just what a cyclone is. Just how bad it can get and exactly what can happen to them if they don't get their asses moving. And then, hearing reason, they will listen. Of course they will listen.

Her laptop is open, and she knows two websites[50] that have exactly the information she needs. Just days ago, preparing for this trip, she visited these sites. And thankfully cut and pasted the content into a word file and saved it, for she has absolutely no 3G reception. She also doesn't have a printer. It will be longhand all the way. She starts writing in a neat print:

Cyclones

A tropical revolving storm is an intense vortex of air swirling in the atmosphere and is named a "Cyclone, Hurricane or Typhoon". It has very strong winds circulating around it in an anti-clockwise direction in the Northern Hemisphere and in a clockwise direction in the Southern Hemisphere.

The swirling moisture-laden energy mass of air and waves is at all times moving in a "curving" motion towards the lower pressure "eye" of the Cyclone. The nearer it is to the eye the higher the waves generated and the faster the wind becomes, and this accelerates to known speeds of up to 180 miles an hour.

In a low pressure system this air is flowing inwards towards its centre, and as it moves closer to its "vortex" the wind speed accelerates. Prior to becoming a tropical Cyclone there are three indicators in the category of a Low.

1 – Low pressure area	less than 17 knots (30 km/h)
2 – Depression	17 to 27 knots (30 to 50 km/h)
3 – Deep depression	28 to 33 knots (50 to 60 km/h)

Once the wind speed increases above 35 knots then the revolving mass begins to form into a Category 1 Cyclone with winds of about 35 to 50 knots. As time passes the barometric pressure drops lower and the winds increase their speed.

In a large Cyclone there may be a barometric drop to 920 millibars or less, and the lower it drops is an indication of the pressure difference between the High and the Low. The lower the millibar reading the greater the pressure difference and the faster the wind speed becomes.

The air mass is rising upwards from the centre of the vortex like an invisible tornado, and the circular wave movement drives floating objects towards its centre, and the nearer the centre, the higher the wind speed and the more dangerous the waves.

Even though a "calm" of negligible wind is experienced by those on land when the centre of the eye of the cyclone passes overhead, this is not the case on the sea, because on the ocean the Cyclone "core" is a deadly mass of very confused and breaking waves that approach the vessel from every direction once the eye has been "entered".

All cyclones cause a high incidence of injury or mortality.

Vessels sometimes attempt to follow the direction of the waves in order to lower volatility. This strategy is ill-advised as all waves are travelling towards the eye in a "roundabout" manner, thus driving the vessel inwards and into greater danger and higher wind speeds.

The outer "edge" of the eye is in fact the point where the wind is revolving around the perimeter of the eye in a completely circular motion at the fastest intensity and highest and steepest wave heights. In a category 4/5 Cyclone wind speeds may reach 180 miles an hour. Vessels should avoid this situation.

While rules of thumb about timing are imprecise, the following can be stated with a medium level of confidence:

***72 hours:** At first there are no apparent signs of a storm. The barometer is steady, winds are negligible to light and variable, and fair-weather cumulus clouds appear... the swell on the ocean surface increases to about six feet and the waves come in every nine seconds. This means that the storm, far over the horizon, is approaching...*

Julia stops momentarily, reads what she's copied from her word file with some edits and additions, and nods to herself. This is good. These are the facts. This is what the others need to know. She'll break it down by the hour and tell them exactly how it's going to play out, and as it unfolds they will see and believe.

***

Outside, dusk is approaching. She's writing again, studiously copying ugly truths from the file on her laptop.

At daybreak, it was just another day. For Sam and Erin, Lorenzo and Angela, it still is. But for Julia and Juan-Franco, the sun is about to set over a world fundamentally different from the one over which it rose fourteen hours before. An unwanted menace has started to announce its inevitable arrival. For those who understand the signs, the inevitability has been more certain with every passing hour. The cyclone is coming.

The cyclone is coming.

It's coming, coming, coming.

But Julia has not convinced the others yet. Not being convinced, they cannot be galvanized into the collective action needed.

And so it is with climate change and biodiversity loss and much else besides.

All the world is a yacht. Every PECCA is an approaching cyclone, surely on its way but visible only in small and subtle changes. And all of humanity is six people on the yacht. Julia and Juan-Franco, who are convinced. And Sam, Erin, Lorenzo and Angela, who are not.

11

ALL THE EVIDENCE AND A SONG

Julia pauses and looks out at the setting sun. The day started so blue and bright, but a high, overcast gloom has stolen in, and now the red fireball that dropped below the horizon the previous evening seems diminished to something less celestial.

She knows that the light will start fading in perhaps twenty minutes and that if she doesn't give them the facts now, the night will pass before she's has another chance. It's all there now in the neat print, three and a half pages. She briefly scans over the last of it again:

30 hours: *The sky is covered by high clouds. The barometer is falling at .1 millibar per hour; winds pick up. The ocean swell continues to increase.*

24 hours: *Small low clouds appear overhead. The barometer is falling by .2 millibars per hour, the wind picks up to 34 mph. The wind-driven waves are covered in white caps, and streaks of foam begin to ride over the surface. Evacuation should be completed and final preparations made by this time. A hurricane warning is issued, and people living in low-lying areas and in mobile homes are ordered to evacuate.*

18 hours: *The low clouds are thicker and bring driving rain squalls with gusty winds. The barometer is falling steadily at half a millibar per hour and the winds are whistling by at 46 mph. It is difficult to stand against the wind.*

12 hours: *The rain squalls are more frequent and the winds don't diminish after they depart. The cloud ceiling is getting lower,* *and the barometer is falling at 1 millibar per hour. The wind is howling at hurricane force at 74 mph. The sea advances with every storm wave that crashes ashore, and there are foam patches.*

***6 hours**: The rain is constant and the 92 mph wind drives it horizontally. The barometer is falling at 1.5 millibars per hour, and the storm surge has advanced above the high tide mark. The sea surface is a whitish mass of spray. It is impossible to stand upright outside without bracing yourself.*

***1 hour**: The rain becomes heavier. The winds are at 104 mph, and the barometer is falling at 2 millibars per hour. The sea is white with foam and streaks. The storm surge has covered coastal roads and 16 foot waves crash into buildings near the shore.*

The eye

Just as the storm reaches its peak, the winds begin to slacken, and the sky starts to brighten. The rain ends abruptly; the clouds break and blue sky is seen. The barometer continues falling at 3 millibars per hour and the storm surge reaches the furthest inland.

The winds fall to near calm, but the air is uncomfortably warm and humid. Huge walls of cloud appear on every side, brilliant white in the sunlight.

At this point, the barometer stops falling and in a moment begins to rise, soon as fast as it had fallen. The winds begin to pick up slightly and the clouds on the far side of the eye wall loom overhead.

After the storm

***1 hour after landfall**: The sky darkens and the winds and rain return just as heavy as they were before the eye. The storm surge begins a slow retreat, but waves continue to crash ashore. The barometer is rising at 2 millibars per hour, and the winds top out at 104 mph.*

***6 hours**: The flooding rains continue, but the winds have diminished to 92 mph. The storm surge is retreating and pulling inland debris out to sea.*

***12 hours**: The rain now comes in squalls, and the winds begin to diminish after each squall passes. The cloud ceiling is rising, as* *is the barometer at 1 millibar per hour. The wind is still howling at near hurricane force at 69 mph, and the ocean is covered with streaks and foam patches. The sea level returns to the high tide mark.*

Note: Sailors should never ever head a vessel in the direction of the cyclone "body", because once you are "running" with the wind or waves you will be unable to turn your vessel into wind without danger of "broaching" and you will be "forced" towards the centre and, as you get nearer the eye the waves will become so steep as to thrust your vessel under. At all times your vessel must be heading away from the eye or anchored facing the waves using a para- anchor to hold it stationary as you wait "safely" for the Cyclone to move away...

Perhaps she could express it better. Add something here or leave out something in another place. But for the most part, any reasonable person would be able to see that this is what is happening and that action is required immediately. She adds a front page and writes:

'CYCLONES: FACTS AND TRUTHS'

in very large letters that she underlines.

She then seals the four pages in a plastic sleeve, fetches a nylon cable tie from the tool box and works the sharp end through the opening at the top of the plastic sleeve. Ready now, she walks out.

Looking at the others, she feels relief that she didn't wait longer. With the sun setting there is a restlessness on the deck. In a while, people will be wanting to go inside, shower, apply after-tan lotions, prepare food.

Aware of their eyes on her, she walks to the mast and encircles it at chest height with the cable tie. For an anxious moment she wonders if it'll be long enough, but it doesn't let her down. She inserts the sharp end into the pawl and pulls it through, tightening it as much as she can.

In that moment her movements are assured, approaching triumphant. A second later she realises that again she didn't plan far enough ahead. Is she moving back inside, waiting for the truth pinned to the mast to speak to the non-believers on the deck, or is she sitting down with them? Before she moves, she already knows that this small detail is the one that will determine whether they read what she has written. She made the entry, but did not plan the exit. She feels the uncertainty enveloping her, and she knows it is evident, and again she feels utterly stranded and ridiculous.

***

In the hours since Julia first came onto the deck, Juan-Franco has been trying to blend in, and remain indistinguishable from the other four. He had a Mochito soon after Erin started the trend, ate some snacks, drank a great deal of water, and had a beer. In between he spent most of his time staring at the horizon. He thought of Esthia. She must know by now. He wonders how she's reacting, what she's thinking.

Again he recalls the brief period when they first met, when she was completely in tune, more so even that him. The time between walking in the rain and being on the ferry in Greece. For that unreal spell she seemed to have supernatural powers. Everything fired in concert for her: mind, emotions, body. She seemed to possess all wisdom, all knowledge, all faith. Everything would be okay. Nothing bad mattered. They lived only in the present and only at full intensity.

His curse is that he cannot forget. After all these years, after all the times of seeing just the shell of what remains of Esthia, he still remembers everything, just the way it happened. The chill of very wet clothes against his body as they reached the end of their walk that day in the rain. Her warm hands on his cheeks as she kissed him and said to go for a cappucino. The inadvertent skip in her step when she approached him, every time he knocked on her door. And the smell of her washed, long hair as he stood at the promenade with his chin on the back of her head and his hands on the rail next to her elbows, and they watched the waves roll in through the mist.

He cannot forget, and he keeps thinking she'll wake up and revert to normal. But he also knows that perhaps it wasn't normal, wasn't the real world, probably never can be again. In many concerts he plays, it is all about finding her. Refreshing his memory about how she was. The moment that happens, the intensity is there, the audience feels it, and it all builds to a crescendo of searching and half-fulfilment and visceral longing. Those shows are all a success.

She must be so worried. He wonders if she still has the Pablo Neruda book, and whether she'll read it now.

The swell is still unusual. If anything, it's higher than before. Perhaps the frequency is higher too, he can't say for sure. He lifts his chin, and on his cheek he feels the first suggestion of a breeze.

It's coming.

As the sun goes down, he watches the others, and feels the stiffness, the contrived relaxation. As if they're a secret society knowing it has a spy in its midst, and as if everyone is scared of being identified as the guilty one. It's because of Sam. They all know he's watching them, like the Inquisition.

Then Julia comes out and ties the plastic sleeve with some kind of information to the mast. Juan-Franco sees her hesitate, not knowing what to do next. He also sees the others looking this way and that – everywhere but at the mast.

He rises.

"Just let me fetch my guitar," he says. "Julia, why don't you keep my towel warm?"

Instrument in hand, he emerges a minute later into silence, knowing that he's secured the grand entry. They're all looking at him, even Sam, who's removed his sunglasses.

"I haven't played this trip," he says as if to himself. He quickly tunes the guitar, then plays a short instrumental. He looks west, sees the sun touching the water, knows the moment has arrived. He has to do it now, while they can see.

"Wow, it's sticky," he says. "Really humid." He gives the guitar to Erin, and starts unbuttoning his shirt. First the right sleeve, then the left. The tricky one that always takes longer because he has to do it with his other hand. He knows every pair of eyes is morbidly fixed on him, the Broken One.

He starts at the bottom and undoes the other buttons. One by one. Going slowly, like a stripper. Then he opens the front of the shirt a little, extending the suspense. And then takes it off, feigning greater disability than he has. Letting the one sleeve get stuck on his arm, and loosening it with some difficulty.

Juan-Franco extends his good arm to Erin to get the guitar, and sees them wide and transfixed. He knows everyone is staring at the grotesque juxtaposition of rippling left and wasting, emaciated, right. The left of a boxer and the right of a hunger striker. At the patchwork of skin grafts and scar tissue around his ribs and back. The long scars and the short, the wide and the narrow. Deep pink to very white to the deep tan on his hands, and many shades in between.

Without introduction he starts to play. Like always, the right hand strums a simple rhythm – that's what he can manage with certainty. And the left does whatever he and the song ask it to do, with utter certainty and precision and at whatever speed required. In this case, it's only a few chords. Juan-

Franco knows the five people on the deck have just been amazed, revolted, shocked, and that for the next while, their attention will waver, then start to return. The song, his own which he performs often and is called *Unbalanced Wheel*, is about a businessman meeting a beggar. It normally only has a short intro, but this time Juan-Franco starts very slowly and speeds up gradually, extends it to two or three minutes, improvising and adding wordless humming in places. He makes sure to turn from time to time as he plays, displaying his body from various angles for everyone to see.

The sun has set now, and darkness is falling quickly, moving their attention back to his head, his face. He concludes the intro, slows down the rhythm again, and begins to sing:

> There's always hope in the beggar's eyes
> always looks up at the
> next guy passing by
> Says: "Man I am down –
> I'm in the gutter
> autumn in the morning,
> rain in the air
>
> Does he know
> that life is an unbalanced wheel?
> I just go,
> I go by the way that I feel
> The way I feel...

Erin looks at him as he improvises between the first verse and the second. It's dark, and she knows nobody will see the tears on her cheeks. She has no thoughts, just a surplus of feelings, an irresistible sense of the poignant. She thinks of her brother and her dog and Julia's warning of earlier in the day and of the swell that seems to be getting higher all the time. And the image if Juan-Franco's disfigured body as he took off his shirt.

Lorenzo has taken Angela's hand, and they steadily move closer to each other. He puts his other arm around her shoulders.

> I've got his dreams
> he's got my freedom
> he sees my suit
> and I envy his smile...

Angela leans the back of her head against Lorenzo's chest and closes her eyes. Juan-Franco's voice sounds hoarse and raw, more so than she noticed before.

Julia too, notices it. She's thankful that the troubadour's intervention has allowed her to blend back in after she was so stranded at the mast, but in the twilight she can hardly see her plastic sleeve against the mast, and she wonders if anyone would still be able to read what she has written, even if they want to.

Sam's jaw is clenched and only the deepening shadows allow him to hide his agitation. He knows that Juan-Franco has all the others in the palm of his hand and that right now, he can steer them any which way. There will be a short moment at the end, Sam knows. A second during which the situation is begging to be defined. That's where he'll have to pounce, or the dawn will see them sailing away from the reefs, away from the island and towards certain financial ruin. All of it while running from a phantom cyclone that never comes.

Juan-Franco changes the rhythm and the key to suggest the bridge and last refrain. He too knows there will be a moment when everything will be defined, sometime really soon, and he knows Sam will be ready. He thinks it over, weighs it, realises that it's too soon to go for broke. He's told his story without words – the story of what a body looks like that's collided with a coral reef, slammed down by a very big wave. He'll let Sam do his thing. The image has been planted, the suggestion made. He'll proceed from there in the morning, when everyone has tossed and turned in their beds and bunks on a gathering swell and tried throughout the small hours to keep shut the gates that imprison an expanding anxiety.

As he sings the last lines, he scans the ocean, but it's become a dark, undifferentiated mass that tells a tale neither of normalcy, nor of doom. The morning might tell, or the breeze, if it continues to strengthen.

Why do we have to carry
all each other's crosses,
live all the other's
unchosen lives?

All I know is
life is an unbalanced wheel
I just go,
I go by the way that I feel

the way I feel…

All I know
Is life is an unbalanced wheel
I just go,
I go by the way that I feel
the way I feel.[51]

Knowing that Sam is about to intervene, Juan-Franco lowers the volume of his strumming considerably, stops singing and starts toying with a few riffs. He gives Sam the opening, but doesn't become part of it.

"The perfect end to a perfect day," Sam says, rising and stretching. "Let's go inside and get something to eat."

He gestures as if in chivalry, and waits for the first movement, fearing that if he goes first, he won't be followed. Angela and Lorenzo move first, then Erin, then Angela. He brings up the rear.

Juan-Franco stays put, sending harmonies filled with patterns and intelligence into the inane darkness surrounding the yacht.

12

Green Gandhi

It's an early, misty morning near Samos. Juan-Franco is awake and smoking, watching Esthia sleep. There's a gentleness in his eyes that's been growing ever since the day they walked 'under the rain'. The ferry sounds its horn and the attractive, young woman stirs. As she opens her eyes, she sees him watching over her, sees how content and self-contained he is. She senses in that moment that he's destined to lead people and to pioneer new things. He'll have a big and rewarding life.

Everything's aligned. They're young and in love, and anything seems possible. For the past weeks, her sense of vulnerability has been gnawing at her. With him, she knows now she'll be closing circles from her youth that she's abandoned mid-stream. Confronting painful memories and navigating cross-cultural complexities. It scares her – having to feel so much, risk so much, synthesise so much.

In the world they would end up living in for the time beyond, she spends the morning uneasily weighing all this, and in the late afternoon her fear gets the better of her and she tells him she can't do it. Not in the flesh. It's too much. His eyes age as she speaks, but retains the gentleness. In time, he becomes a performing artist. His voice is hoarser than she remembers from that day he sang her all the verses and refrains of *Vincent*, a capella. He's always energised by her, she invigorates his art. He has the big life she saw for him, and he leads people and pioneers things. Increasingly from the stage and not from an office. And yet he's unbalanced, like his left and right sides after he dived in to save her and got smashed on the reef. He's often alone. They meet once in two or three years for a walk or a cup of coffee and it's as if none of the women he meets in between manages to change him. The next time, he's the same Juan-Franco again, with same gentleness that for her sake she is delighted to see, and for his sake sad that he hasn't lost. But they're always together, in a world just beyond reach. The certainty makes the world they actually inhabit somewhat tolerable.

In some parallel world that they can both sense but not reach, she spends that same morning on the ferry weighing her options, but the look in his eyes, as he watched over her as she slept, outweighs everything else. The challenges are the same, as daunting. But the big life she sees is their big life, the pioneering work will be their joint effort. By the late afternoon, they stand at the rail and watch the sun go down. She knows he's been asking in his quiet way. With that certainty she had as they walked in the rain, she kisses him and then brings his ear to her mouth as she whispers: "Okay, let's go for it." They're always together after that, her children are his children. She suffers through many cross-cultural challenges and slowly starts to close some circles that she once thought were abandoned. Juan-Franco is a successful social entrepreneur. He's balanced and never has to dive off a boat to save her. Sometimes, on a Sunday evening, he takes out the guitar that he built himself as a youth and plays a few songs. He does it well, but it's other people's songs and the content stays some distance away from him. Never becomes him, or him it.

They ended up in the other world, but the one that could have been sustains them.

Now, the deck is abandoned. Just Juan-Franco with his guitar, playing to the silhouetted outlines of the yacht and the stars, shining brighter all the time. He thinks about Julia's precious knowledge, still stuck on the mast, remembers how once he became Green Sysiphus as he tried to help the Small Island States. He finally despaired and started looking for other ways. Knowing Sysiphus very well, in time he made two other friends whom he would consult about the best solution to an environmental impasse. The one is Gandhi,[52] the other Bruce Willis. Within the context, Green Gandhi and Green Die Hard. The one preaching passive, peaceful resistance, the other violence.

***

Erin lies in her bunk, one of two on the yacht, closer to the bow than the two double suites. Feeling the swell go up, and down. She's starting to feel queasy.

It's so hot. She wonders if such humidity is normally associated with cyclones. She could sneak over to the mast and try to see what Julia's document says, but it's too dark now to see anything. She listens for strong wind, but there's nothing to concern or console her. Just the vague queasiness and the nagging doubt that has caused her to put away the novel she was reading.

She thinks of her parents, back in Manchester. They always wanted to help protesters. Dissenters. People like Julia.

When she was a teenager, there was talk about a second runway at the local airport. Her friends didn't care, and she didn't care. Gabrielle had a new single on the charts and they talked about wearing an eye-patch like her when they went to the next party.

But her parents cared. They belonged to Friends of the Earth and said the new runway was wrong and that it would cause a lot of harm. For weeks, they went to secret meetings every few days. In their conversations at home, the name 'Swampy' came up more than once. Her father wanted to be part of something that was being planned, but her mother worried he might lose his job if he stayed away so long. They spoke into the small hours one night when they thought the kids were sleeping. She heard 'Swampy' again and 'direct action' and 'making a stand'.

Just days later, it was all on TV. Swampy and his friends had occupied the area where the second runway was to be built. Most of them were in tree houses or tunnels and some were rumoured to have encased their legs in cement to make it harder to dislodge them. Apparently it was all illegal and the sheriff would be removing them, one by one.

Suddenly the pop charts and Gabrielle and her eye-patch were forgotten, and Swampy was the hero at school. They didn't really understand what was so bad about another runway, but the illegal part was cool, and defying the security guards, and challenging the sheriff to come and get them out – if he could. Some kids had a music tape made by the protesters and they played it at school during the break. It had two guitars and people singing songs written at their camp in the trees, and under the ground.

Every afternoon when Erin's dad came home from work, her parents would get into the car and drive to the airport with food, cigarettes and newspapers, sometimes blankets. The second time, she said she wanted to join them, and they all got into her father's panel van and drove there.

"The security guards work for the airport and have no power to evict anyone," her father reassured her as they approached. "and the sheriff knocks off at five. So between now and sunset all the friends and family of the protesters come here to the fence at the back of the airport to lend moral support and bring new supplies."

He added something about logistics being the making or losing of every war, but it was lost on her.

They got out of the car and walked closer, supplies in hand. There were perhaps 30 people on their side of the fence, and maybe 15 on the inside.

"How was your day?" they'd ask each other like a family member asks when someone comes home from work. "Okay thanks, how was yours?"

All so normal, even mundane. And yet they all had this utter certainty that what they were doing was right. That they were the good guys and that the sheriff and the authorities who wanted to build the runway were the enemy.

"We have to hurt them," Erin heard her father say to a guy with a bandanna for whom he'd brought cigarettes. As if he was on the inside, one of them. "Make it so expensive to get us out that they won't ever oppose us again, anywhere else."

The guy nodded as he lit a smoke. Erin wondered again what it was all about. It felt like her parents thought she would instinctively know, that any decent human being would just know. She didn't really want to ask. And her friends didn't really know either. They said different things about capitalism and greed and exploitation and thought it had something to do with being 'green'.

Then the sun started going down behind the concrete of the airport buildings, creating odd-shaped, imposing silhouettes. A while later the security guards with their neon green jackets were still visible, with almost everything else hidden by deep shadow. Everyone said goodbye. The protesters walked back to their trees and their underground caves and Erin got into the car with her parents to drive home and switch on the television to see if the protest was on the news, as it had been two nights before.

Life went on. Every day some protesters were removed, some screaming and kicking, others quietly allowing themselves to be carried away. The story featured in the local paper every Friday and on the radio almost every day. By the end of May some girl nicknamed 'The Worm' was dislodged from her encasement and the noose around her neck, and arrested. Two weeks later the last protestor was gone and the construction could begin.

But in the autumn and into the winter a girl named Butterfly in America was all over the news for sitting forever in the top of a very tall Redwood tree to prevent it from being felled by loggers. Erin's parents followed the matter with great interest and wore T-shirts saying "CAN YOU Spare Some CHANGE?"[53]

While this happened, the construction on the runway continued and some of Erin's friends who dropped out of school were being employed there. They had these jobs for three more years, until the project was completed.

Other things became news, and Erin and her friends discussed different topics. She never quite understood what the airport thing had been all about for her parents. Just that they liked people who had the conviction to make a stand. Like Julia did today, a little bit.

Erin feels the humidity again, and wonders if it's worse than an hour before. Maybe if she'd understood about the airport she'd have known more about who is right and wrong now. Would have known what to do.

Juan-Franco's disfigured upper body flashes before her, and the refrain of his song, "life is an unbalanced wheel..."

Listening to him, it felt like he knew who was right and wrong about the cyclone. Knew for sure. But he wasn't saying, for some reason.

Unlike Sam.

Sam is so sure. But there's something about him. She doesn't trust him.

But he's sure that everything is okay. There's no reason not to trust him. She should trust him.

Everything will be okay. Time to sleep.

She turns over, adjusts the pillow, senses some lingering doubts circling persistently like vultures above a carcass.

Time to sleep.

Everything will be okay.

***

Juan-Franco is still playing his guitar, allowing his hands to go where they want to and do as they wish. It's an old habit, and it allows him to think – perhaps more than doing nothing would have done. The guitar provides a focal point for the conscious mind, and yet requires no deliberate thinking, because he isn't trying to achieve anything. The mind and the imagination can wander.

Where Esthia is, it must be a cold afternoon. Does she know about the cyclone and the fact that they may be reaching the end of their history together, the saga of being apart in the world they know, and together in the one they believe in?

Juan-Franco realises that he has to visit Gandhi. He's started playing the aces of persuasion and in the next 12 hours, will play a few more. Perhaps complete unanimity will come to the yacht before it's too late and they'll change course. Or perhaps there will be a time when only Green Sysiphus would try again, fruitlessly, to roll that boulder over the top of that hill. Then, Juan-Franco would be down to Green Gandhi as the last option before only Green Die Hard is left standing. The way of Bruce Willis, of the sword and of the whip. He has to get Gandhi's view. What can be done peacefully?

It will be a journey only of the imagination, with his body staying behind on the deck, fiddling with the guitar. This makes many things possible, and he toys with the idea of taking Esthia along. The two of them off to India together, like they once went to Samos. Just them.

Moments later, he maintains the status quo of many years, discards the idea and goes alone. He finds the Mahatma wearing *dhoti*, sitting on a rock on a high mountain, overlooking a steep drop and then an immense plain below, stretching to the horizon. "I would offer you peace, and love, and friendship", says Gandhi, without turning round. Juan-Franco looks at the back of the frail sage, and beyond him to the vastness of the plain. They see the same expanse, and they have to, to have a conversation about it. "But you're not?", Juan-Franco asks. The Mahatma is quiet, and it's the visitor who continues:

"Many people have started to follow your principles to protect nature and animals."[54] he says. "They choose non-violence. They chain themselves to trees about to be cut down. They break into laboratories and set free the animals on whom human products are tested.[55] Some paint graffiti slogans on the expensive symbols of consumerism. They always own up to what they've done, and agree to be prosecuted and punished..."[56]

Juan-Franco pauses to hear Gandhi's response. "Without action, you aren't going anywhere." Gandhi says after a pause, ostensibly in agreement.[57] But the discussion is only beginning.

"In a gentle way, the world can be shaken." Juan-Franco continues. "Even a song can change everything. Displaying an injured body."

Quiet descends as Gandhi waits to see if more will come from the one who seeks his wisdom. A breeze kicks up and drives the balmy air from the floor of the plain, up to where they are, Gandhi sitting, Juan-Franco standing. It moves their clothing and Juan-Franco's hair, then subsides.

"Strength does not come from physical capacity.", remarks Gandhi. "It comes from an indomitable will."

"Time isn't my friend." Juan-Franco responds. "The wrong hands are on the tiller of my world, and doom is around the corner. A storm is coming. I can't change the course of the vessel just by refusing to collaborate."

Gandhi is quiet for a long time as both men look out over the limitless plain. Juan-Franco realises that they're looking at the future, stretching away to infinity.

"You're already the change you want in your world", he hears the old man saying. "That much you've done, I can hear it. You must keep trying the gentle persuasion, until the very last opportunity. But when the time comes, I would risk violence a thousand times rather than risk emasculation ... I do believe that where there is only a choice between cowardice and violence, I would advise violence.[58] I would rather have you and your comrades resort to arms in order to defend your future than that you would, in a cowardly manner, become or remain helpless witnesses to your collective dishonour."

This time it's Juan-Franco who remains silent, hoping for more. When it comes, it's wistful, especially coming from such a wise and ascetic man: "The world is not entirely governed by logic", Gandhi says. "Life itself involves some kind of violence and we have to choose the path of least violence."

He's quiet then. For a moment longer Juan-Franco and Mahatma Gandhi look out over the future, in silence. The next moment, the journey's over, and the guitarist is back on the deck of the yacht. He thinks about the words he just heard, of their wisdom and yet their realism.

The sea is restless, the swell seems to be growing, the wind is getting up.

Some things can start now. Right now. Like building the sea anchor. Two people can make it. Slower, yes. In three times the time six people would take. But they can do it. One person can also make it, even with one strong and healthy ar m, and one that is damaged and weak. After a long discussion on the deck and after a somewhat risky swim with Julia, this is exactly what Juan-Franco will do.

***

Julia lies in her king-sized bed in the luxury cabin she shares with Sam. Sam is quiet and motionless, probably sleeping. She didn't feel much like food and came to lie down straight away. She feels the vessel rise on the swell, then drop, then rise, then drop. Would they read her three and a half pages in the morning? She wonders what else she can do. Three times she has played her supposed trump card, and three times she's been met by indifference.

What can she do differently? She pictures the plastic sleeve filled with knowledge. Like an unopened capsule filled with signs of civilization, sent on a spaceship to outer space. She needs to get their attention. Like Juan-Franco did when he took off his shirt. She wonders why he did it. She was half-annoyed at the time, for she felt he was drawing attention away from her plastic sleeve. Her exposition. Her communication of knowledge. But perhaps he already knew they weren't going to pay attention to it. Perhaps he already understood that something else was needed to draw their interest.

She hears him playing riffs on the deck, humming along from time to time. She needs to talk to him. Ask him what they should do.

She gets up to join him.

13

Overboard

Juan-Franco is playing the opening notes of *Stairway to Heaven* as Julia emerges onto the deck. His right hand lets him down with the picking, and he hits two strings when he was aiming for one. He stops and starts again. Julia waits for a mistake, but it doesn't come, and he starts humming:

> There's a lady who's sure
> All that glitters is gold
> And she's buying
> the stairway to heaven...

His fingers find their way to another tune. She sits down flat on the deck, legs crossed, close to his feet. He continues playing and she understands that it's a tribute, a salute, that she's entered the space of Juan-Franco and his guitar and that he's chosen to let her share it with him.

She thinks of Sam, probably in a slumber already, only a few steps away. And yet in a completely different world. Just the previous night she would have felt awkward sitting on the deck like this with Juan-Franco. So intimate, with the breeze in their hair and the occasional sound of the swell colliding with the hull of the yacht. His tunes, the fragments of songs. Some his own, some well-known. And the dark, beyond the little circle of light emanating from inside, the utter dark. Not a single star visible behind the cloud cover that became more pronounced as the day wound down.

The radio report has changed everything. Divided believers and non-believers, wheat and chaff, the important and the frivolous. Just a day ago she would have worried about Sam's reaction at her sitting here.

"Is he sleeping?" Juan-Franco asks without interrupting his fingers, again attempting a tune too hard to play in a performance.

"Guess so," she replies.

He plays on, and switches soon to the song he played when the sun was setting, but attempts some further finger picking.

"Is it still deteriorating?" she asks.

"The nerve damage? Not really. If I exercise and practise a lot, it sometimes improves. But mostly it stays as it is."

"And you feel nothing that side?"

"Not nothing. And just not in the upper body. It's like a dead leg when you've sat in a funny position and blocked the circulation. Just before the pins and needles start."

She sits quietly for another fragment and a half, realises she needs to understand how close she can get.

"It's still you and Esthia."

It's a statement of fact, derived from the perception of an immediate, undeniable reality. Like noticing that the night is dark. That this is a yacht. Like feeling the breeze picking up.

Juan-Franco and Esthia.

He plays on. It wasn't a question.

"You've sacrificed so much," she says, knowing immediately she's just chosen the wrong words. He doesn't reply immediately, but his hands slow down to a snail's pace, as if he's thinking it over. Then he puts down the guitar, and sits down opposite her on the deck. Their legs touch, and he makes no effort to move away.

"She made me an artist," he says at length, as if he's figuring it out for himself for the first time. "Artists need an inexhaustible source of inspiration, or they burn out. Or they lose their edge, their legitimacy."

"And so you think this is the way it was meant to be?"

"This is the way it is. There is good in it."

"You never wish for more? You just live in the moment?"

She was sure he'd agree, but he doesn't. His voice is hoarse, pensive: "I always wish for more. I always remember moments when there was more."

Julia sees him move before he goes on: "...and before I can forget, in that instant when I sense it fading like a torch with old batteries, I go back and relive it again. Every fine detail. I write a song about it, and the memory is sharp again, crystallized. Perhaps after these years what I have is not a memory, but a memory of a memory of a memory. But I remember. And while I remember, I am never fully in the moment."

She closes her eyes, as if to test the unreality she feels. When she opens them, he's watching her.

“I’m from the other side,” she muses. “I’ve always thought I was happy. I’ve always been in the moment. With my first husband. With Sam, again, with all the relationships since school. In the moment, there’s nothing to compare with, and so whatever I have is good. Once you have a child, it becomes even easier. You’re too scared to do anything that allows comparison.”

“And now? After today?”

“Now I know there’s a gorge between me and him. Now I know I’ve never been in love. And that I’ve never been hurt.”

He turns his head to the side, stares into the dark. “Falling in love is realising the relationship could hurt us more than we’ve ever been hurt before.”

The breeze gusts through the rigging, as if to utter words of warning about a change in weather, then subsides again.

Julia thinks of her mother, of the syringe and the needle and the ambulance. Of the swallow. A conversation for another day, if there will be one.

“Do you think I’m serious?” she then asks. “Academic and scientific? Impassionate?”

He looks at her, waiting, as if he knows there’s more to come. She relents. “On a day like today, I have to wonder. With Esthia and all that happened, it feels like you could die any day and it’d be okay. Like the most important thing in your life is in the past. But for me, yes I have a daughter. And yet it feels like the most important thing never happened. Like there’s more reason for me than for you. More reason why I can’t die tomorrow or the day after. Not yet.”

She notices his good hand stroking the guitar, feeling the strings like it’s always feeling something, and regrets the honesty of her words. “It’s okay,” he reassures, and she can just see the white of his teeth as he smiles, or grins. “I have my work. And it doesn’t look like it’ll ever be done. So there’s good reason for me to survive.”

“Nice try”, she feels like saying at his deflection of the issue. But why shouldn’t she? The cyclone is coming.

“Nice try,” says Julia.

“If I could see you now,”, he starts, “I’m sure your face would be changing. Perhaps a deeper gloss on your cheeks. Perhaps a reddening around your throat. And I’d think you’re angry, but trying to be civil. So no, I don’t think you’re impassionate.”

His voice is gently teasing, but it’s what’s needed in this strangest, most surreal of moments.

She's still wound up, but aware of his kindness. Her voice becomes softer. "And so you go through life sustained by memories.?"

"Not just memories. There are women. And friends."

"Sustained by memories and groupies," she teases back.

He agrees, tacitly, and picks up the guitar again, fools around with it. He sings a verse from *Suzanne* by Leonard Cohen, as if to illustrate the point.

"Let's go for a swim", Juan-Franco says.

Her first reactions are thinking of her daughter, and mild shock. The swell must be eight feet or more. But she forces herself to take the idea, sit it down next to her, look it over, make friends with it. There could be currents, sharks, whatever. But the cyclone is coming, perhaps 36 hours from now. And he'd have a reason for asking.

"Explain, please?"

"Well, why do you want to survive the cyclone? To do fun things. So do them now. Right now. Don't get so wrapped up in it that you forget to get around to the good stuff."

She thinks of his work. The advocacy, the activism, the campaigns and speeches. The sustainability projects. And then the music and the stories. On the face of it, a different, unrelated pursuit. But suddenly she grasps that the one cannot be without the other. That it's all the same thing. Warning people about a cyclone that is coming. Trying to miss it.

"Okay," she agrees. "If we can tie a lifebuoy to a rope for me to hang onto. I'd feel sort of silly at the thought of drowning and then Sam survives the cyclone and has to tell Stella what a daft and irresponsible mother she had."

"Groupie," he teases again.

She knows it's a compliment of sorts, an encouragement to take another step on a long road. And smiles.

They jump together soon afterwards, arms around the lifebuoy. The water is cold at first, almost shockingly so after the stifling humidity, but as they surface it's already becoming comfortable.

Julia defaults, as always, to the scientific and the factual, and feels disbelief at the height of the swell as they bob up and down.

"This isn't eight feet!" she gasps through the wet hair on her face. "This is twelve at least!"

She counts slowly, the seconds between the swells, and gets to eight. Eight seconds. Counts again. Seven. Counts again. Eight. Then eight. Then nine, then eight.

"It's close," she says to Juan-Franco, adjacent to her as he hangs onto the other side of the buoy. "Much closer than I thought. Perhaps 12 hours closer."

He lets go completely, disappears, returns.

"No currents," he claims.

"On the contrary," she corrects him. "One huge current. One huge pull. You see how the swell runs in the same direction the yacht is facing? We're being pulled into the eye. Everything goes there. But everything moves together, and so it feels like there's nothing."

Without further explanation he loosens her grip on the buoy, and pulls her a few feet away. She's just explained to him why it's safe, but she still gets a fright. Still she wants to stretch and grab the buoy and hang on.

Exercising all willpower, she refrains and treads water, noticing that the distance to the buoy is remaining about the same.

"This is what makes it worthwhile for me," he says. "All the struggle to save this and conserve that. Experiencing sometimes the fullness of what is here, now. Knowing it could slip away any moment. Or I could."

He swims back to the buoy, then extends a hand, pulls her closer so she's hanging on again. They bob up and down as they did before. Looking at it, just for the moment it's only a swell, not the precursor of anything. And it feels to Julia as if she's being rocked to a lullaby. Closing her eyes, she imagines the stars as they must be above the cloud cover: bright and reassuring in the way they're ordered – into the same constellations as ever.

She looks at the yacht, where some lights have been extinguished, and the perspective of not being on it creates the illusion that she's not part of it. Not its past, nor its present, nor its future. And that away from the yacht there is a oneness, a benevolent harmony, gently enveloping her.

14

Blackbeard's Ghost – Disney's accidental master metaphor for human irrationality

There was a spell, and a pirate cast into limbo until such time as he'd done something good.

> Edward Teach, sometimes known as Captain Blackbeard,
> When you come to die
> May your body and soul
> Be racked between this world and the next
> Always to be alone
> May this curse hold fast and true
> May you dwell for ever more in limbo
> Until such time as there be found in you
> Most wicked of all villains
> Some spark of human goodness!

Then, Steve Walker going to Godolphin as the new track and field coach of the local college. The discovery by accident of a book of spells, and Walker uttering the words used to bring someone back from limbo:

> *"KREE KRU VERGO GEBBA KATTO KREE!"*

And the appearance of Captain Blackbeard – wild beard, pirate's hat, ever-present sword and ubiquitous bottle of rum. Only Walker can hear him and see him. To all other people he is invisible and inaudible. Only at the end can everyone see and hear Blackbeard. But he has a fundamental role in everything that happens in the film; He dispenses black eyes, trips up opposing athletes, changes outcomes on the roulette wheel. He saves the inn and the Daughters of the Buccaneers. He defeats Silky Seymour.

There had been Ignaz Semmelweiz. He too could see and hear what nobody else could see or hear: The medical personnel were bringing death to the pregnant mothers. It was carried on the students' hands, after doing autopsies. Conducting an autopsy and then delivering a baby could lead to death of the mother. But the answer was simple – the medical personnel only had to wash their hands in between, and the problem would be solved. But for years and years, this obvious fact could not be swallowed by the medical fraternity. Semmelweiz's truth, to everyone else, was Blackbeard's Ghost.

There is Julia, back on the yacht. Three times now she's called the attention of everyone on the yacht to the compelling evidence pointing to an approaching cyclone. First, she heard it on the radio. When the radio reception lapsed, she pointed out that the weather conditions coincided with the early signs of an approaching cyclone. She even wrote a mini treatise on cyclones by hand and tied it to the mast of the yacht – but nobody read it. She and Juan-Franco see an approaching cyclone, but to all the others the cyclone is Blackbeard's Ghost. Invisible, inaudible, unnoticeable.

There are the experts and scientists working in fields like climate change and biodiversity loss. Telling us all the time that things are going wrong. That, in a manner of speaking, there are cyclones coming. They write reports that they tie to the mast, but most of us don't believe them. What they see, to us, is Blackbeard's Ghost.

Is this coincidence? Or is it human nature?

At the swings in the park are two young friends, Sally and Tom. Sally is small and wearing a green dress. Tom is plump with a striped shirt, red cheeks and very straight hair. . They play on the swings, watched by Tom's mother. She's wearing a brown blouse and purple pants and has the same red cheeks as her son, also the same body shape. She keeps an eye on them and reads a magazine. Moments ago, she gave each of them a little shove to get them going.

They swing, forward, back. Sally notices the cooling breeze on her cheek every time she swings down.

Up, down, higher, faster. Tom, elated at the freedom, also feels the wind on his neck every time he swings down. It's just below his hairline and above his collar. And on the back of his ears.

When they tire of this, they call Tom's mother over to help them off again, and they run over to the seesaw. Sally sits down on the side touching the ground. Tom reaches up with both hands and pulls down on the other end to try and bring it down to ground level, but he's unable to. His mother watches him going to the middle of the seesaw and straddle it.

Balancing carefully, Tom starts shifting backwards towards his end. She waits for the seesaw to tip, but it doesn't, not even when Tom sits at the very end, in the seat.

"Push, Sally!" she hears him urge his friend, who puts her feet on the ground and tries to push herself off the ground, and Tom towards it.

Nothing moves. Tom's mother sees the children turn to her for help. She walks to Tom's side and pulls down gently, expecting an immediate response. When nothing happens, she pulls harder, still to no avail. Eventually she lifts her feet and hangs on the seesaw with all her weight, but it doesn't move an inch.

It seems impossible.

But the reason is simple: Blackbeard's Ghost. Invisible to all. But profoundly affecting everything.

The same happens every time Julia tries to convince Sam and Erin, Angela, and Lorenzo.

And so it often is when science tries to communicate about something controversial relating to public policy. There is an invisible weight on the one side of the seesaw. The ostensibly lighter weight stays down, and the weight of evidence stays up.

Change does not come, despite an approaching cyclone. Things stay the same.

There is a truth. It is cold and hard. Ugly.

And it is this: Human nature prevents us from hearing what science has to say if the news isn't good. Science in the field of psychology knows this. Has proven this time and again, over and over. Accepts this. And yet science more broadly ignores this. Continues to communicate in a way that cannot be heard. Because scientists believe we should all be perfectly rational and should be able to 'hear' cold, logical truth.

Better to perish than to concede. Better that our civilization goes extinct in the glorious pursuit of perfect rationality and reason than survive in the acknowledgment that perfect reason and rationality, with their zeros and ones, are best left to computers. Not to humans.

***

Sigmund Freud reduced a range of observations to a simple truism: Humans defend against uncomfortable and inconvenient truths. Said differently, happiness is a higher norm than truth. We'd rather bend the truth and be happy than face the truth and be sad.

Today, cognitive psychologists generally accept this. They tell us that we see what we want to. We hear what we can live with. We block out the rest.

Irrationality is the norm. Rationality is the exception. Humans are capable of rationality and occasionally succumb to it. Only occasionally.

200 miles from the yacht with Julia and Juan-Franco, Sam and three other people aboard, many community members are still lost in the devastated coastal village, where calm has returned in the aftermath of cyclone Fifi.

On the beach lies the body of a fisherman who was at sea when it struck, it washed ashore 36 hours later. It's bloated and with large eyes that seem to have seen something terrifying just before they froze. The police are removing the body and have stipulated 'cyclone' as the cause of death.

The dead fisherman's brother always fished with him. Was seen going out with him. Never came back. The parents of the brothers have been informed that the brother is presumed dead. The mother is distraught, but the father is unperturbed. "I just know he's alive," he keeps saying. He tells his wife not to change anything in the missing son's bedroom. He 'just knows' that one day, his son is coming home.

The father is in denial. But whole societies do this. When HIV/Aids was identified as the probable cause of increasing, inexplicable deaths in the gay community in the late eighties in the USA, did everyone say "Right, now we know the problem, what do we do about it?"

No. The truth is too inconvenient. Too painful. Many people could die. Lifestyles would have to change, blood banks would have to totally re-invent their safety procedures at great cost, and the government would be left to pick up the pieces.

Initially, the US government, the gay community, the blood banks and the scientific community all refused to believe the hypothesis. Scientists searched hard for other explanations that would have less severe implications. They investigated, for instance, whether drugs might be at the root of the problem. A conspiracy theory was hatched by members of the gay community, claiming that the whole thing was right-wing scaremongering intended to undermine the gay lifestyle.

And the band played on,[59] like it did for Ignaz Semmelweiz. And people died. And the band played on. And people died.

Until denial was no longer possible.

Denial, unlike diamonds, may not be forever. But there are many other defence mechanisms that can be employed when denial runs out.[60]

None so deaf as those who will not hear.

***

Julia cannot recall exactly what she said when she first ran out. What is however known about the cyclone is that it is life-threatening, strong, and that there is a degree of uncertainty about whether it will hit the yacht. What if she said that there is a degree of uncertainty, it is strong and it is life-threatening? In the former instance, 'life-threatening' is mentioned first and becomes the context, with 'uncertain' becoming detail. We're more likely to react like Juan-Franco than we are to react like Sam. But in the second version, the context is 'uncertainty' and 'life-threatening' becomes detail. We're more likely to react like Sam.

Scientific experiments have shown that we do not treat this information the same as the situation where what is known is merely communicated in a different order, being that there is uncertainty about whether the cyclone will hit them and that it is strong and life threatening.

The first word describing the cyclone is more important than the last. If 'uncertainty" appears first and 'life threatening' last, we're more likely to do nothing than if the words are positioned the other way around. But a computer, knowing only zeros and ones, treats the information the same, either way.

This human idiosyncrasy is just one example of confirmation bias. We have beliefs. We form theories. We look for things that confirm them, not things that force us to reconsider and perhaps change them. And even if it goes wrong, we refuse to budge.[61]

In Miguel Cervantes' classic novel *Don Quixote* (1605), the protagonist is a middle-aged bookworm far removed from being an action hero. But the many books he has read about valiant knights performing brave deeds of heroism have fired him up so much that he sees himself as a knight, riding to the rescue of the powerless and downtrodden. And so, although his steed Rocinante is thin and sickly and his squire Sancho Panza is a simple peasant, he sets out to do good in the world. His delusions know no end. When they approach windmills used to drive grain mills, Don Quixote convinces himself that they are evil giants that he needs to attack forthwith.

Charging on Rocinante, he shouts, "...for this is righteous warfare, and it is a great service to God to remove so evil a breed from the face of the earth!"[62]

In vain Sancho Panza, panicking a bit, exclaims: "Look, your grace ... those things that appear over there aren't giants but windmills, and what looks like their arms are the sails that are turned by the wind and make the grindstone move."

But Don Quixote races undeterred into battle, shouting, "Flee not, cowards and base creatures, for it is a single knight that attacks you!"

He ends up on the ground, black and blue, with a broken lance, positive that by magic the giants had been turned into windmills just as he attacked.

Cervantes understood more than 400 years ago what science will not accept: people are often not rational. We believe what we want to believe.

***

Sam wants to get to the island to sign the contract. He needs to believe that this is possible, because his financial wellbeing and happiness depend on it. Moreover, he believes that business people generally know best about life. That artists are 'iffy' and 'airy-fairy'. That scientists are a bit autistic, unable to connect the dots. The markets know best. Matters regulate themselves through the magic, invisible hand of the market, and the less artists and scientists and politicians interfere, the better things go. In his mind, certainly business people are like parents and other people like children. Business people, like himself, understand the reality of the markets and thus have to think for those who cannot properly do so.

Knowing this, we could have predicted how he'd react to Julia's warning about the impending cyclone: To start, he'd assume that Julia doesn't have the ability to come to a proper conclusion about the information at the radio. If he hasn't been there himself, he doesn't trust that what she says does indeed reflect the truth. She's a scientist. Somewhat autistic. There can't be a cyclone coming, because he has to get to the island. So Julia has to be mistaken. The report must be talking about a place somewhere else.

Pre-believing all this, and unlike a computer, Sam interprets the information that comes his way in a manner that will enable him to keep believing what he already believes and what he wants to believe – that he's on his way to the island and that the weather will permit this. And so, when the swell is high as often it is before a cyclone, he looks at the perfect weather otherwise and says: "It's a perfect day with high swell. There isn't a cyclone coming."

Furthermore, Sam believes that conversation about the cyclone or its very existence should be kept to a minimum. He knows what to do, and the last thing he's going to engineer is to get involved in a discussion with five semi-morons about the best way forward. Information should be managed, filtered. The others cannot be trusted to come to logical conclusions given the facts, and therefore it is best if they are given as little information as possible. For this reason he places himself close to them as often as he can, so he can prevent and divert discussion. Prevent doubt. Keep them on track.

While John Lennon might think Sam is despicable and other liberals might consider him a dinosaur, cognitive psychologists would regard Sam as normal up to this point. If he should pull out a gun to ensure the yacht goes where he wants it to, that would push him into deviance. But that's what he does with his convictions, not how he holds them or formulates them.

This phenomenon has been proven extensively in experiments,[63] amongst other things to prove the irrationality of cognitive psychologists themselves and even jurors[64] and Nobel Prize in physics winners.[65] It's been shown that when opposing politicians talk an equal amount of nonsense, we'll find ways to make our preferred candidate look coherent while leaving the one we don't like marooned on the sandbank of internal contradiction.[66]

***

But how do we do this? How do we make a highly sophisticated brain (capable of matching and even surpassing a computer) spit out wrong answers that we like? If this brain has the ability to find the right answer, how can it fail to do so? How can it sometimes react rationally, and sometimes not?

With computers, its garbage in, garbage out. Or sense in, sense out. No discretion in the middle. Input accuracy is everything. Output accuracy is more or less assured.

Not so with humans.

In political science, a long-standing body of research on "partisan" biases in political judgment (e.g. Taber, Lodge, & Glathar, 2001; Campbell & Converse, 1960) points to another set of limits to rational judgment imposed by emotion-biased or motivated reasoning ... Motivated reasoning can be viewed as a form of implicit affect regulation in which the brain converges on solutions that minimize negative and maximize positive affect states ... Freud (1933) described such processes decades ago, using the term "defence" to denote the processes by which people can adjust their cognitions to avoid aversive feelings such as anxiety and guilt.

In the experiment, perhaps not surprisingly, the subjects overwhelmingly said that the opposition candidate had made contradictory statements. The subjects were assessed while being monitored with an MRI scanner. What is utterly remarkable is that especially when assessing the contradictions of their favoured candidate, emotional loci in the brain that have nothing to do with reasoning were activated. The conclusion drawn from the experiment was that the subjects were actively (and emotionally) blocking their own recognition of their candidate's inconsistencies. It seems that there's a part of the brain that censors the conclusions we're allowed to come to. Stronger in some people, weaker in others. And that even if something slips through that is particularly uncomfortable, our memories can be tampered with in order to restore emotional comfort. We'll forget it quicker than someone who felt happy about what came to mind.

In the time of Joseph Stalin, his opponents were often 'erased from history' after their death. One example is Nikolai Yezhov, who had a distinguished career in Russia and was once photographed with Stalin. Shoulder to shoulder. Comrade and confidante. He was instrumental in Stalin's brutal purges in which hundreds of thousands of people were killed. Yezhov himself was ultimately killed and erased from the photo where he had appeared with Stalin.[67]

Yeshov had once been there, been part of so much that happened. And then something changed. The facts became inconvenient, discomfiting. So they had to be amended, adjusted and re-stated.

Photo: Wikipedia

History was amended, and presumably Stalin felt much better. The man he had executed was no longer a former confidante who'd once stood with him alongside the Moscow canal. And with the photograph proving otherwise no longer around, Stalin could presumably, in his own mind and in the records of the USSR, completely re-write the story of his close association with Nikolai Yeshov – he could do so unrestrained by objective proof.

We humans do this all the time in our own minds – some more so than others. Experiments prove it.[68]

We remember what we want to remember. What we're able to remember while surviving. If there's a very strong emotional reason for it, we can even remember things as they were not.

***

So we have defence mechanisms that essentially boil down to finding a way that makes it possible for us to live in a self-distorted and less painful reality.

Why is this so important?

If Lorenzo grew up in an arid area and hates to see water wasted, and if someone has left a tap open and the water is gushing out into the gutter, does he have to concern himself with their state of mind? If the person who was supposed to close the tap has convinced himself that he has in fact closed it – even though he forgot – does it matter?

For accountability and future prevention, yes. But in terms of stopping the unwanted situation, no. Lorenzo can walk over and close the tap.

But for Julia and Juan-Franco, the state of mind of the other four people on the yacht is critical. Their challenge is persuasion towards positive, collective action.

And this is also the challenge for people who are fighting the new serial killer. The PECCA.[69]

In the terrain of public policy.[70]

If they believe in human rationality, in peace and in love, in harmony between people and in the essential goodness of humans, they will continue to try to build consensus. Hold discussions. Give people time to come around. Look for better ways to communicate and to overcome divides. More than anything else, they'll work towards evidence-based decision making. Trying to take the emotion out by just letting the evidence speak for itself. Letting the science speak for itself.

Perhaps the ugliest of truths about human rationality is that this doesn't work when the matter is complex.

A day before on the yacht, they were all friends, sailing towards an island. A report came in about an impending danger. There was evidence and there was uncertainty. They all knew everything about what was certain and what was not. Julia and Juan-Franco drew one conclusion, Sam another. And one day later, their solidarity has splintered. Julia and Sam will only sleep in the same bed once more, and away from each other, towards the edges. Sam and Juan-Franco both know there is a confrontation coming. That in fact, it has already started. That one of them will prevail and the other will perish.

The evidence has not brought them together but divided them. This is because even previously, they differed in most senses of the word. They had different things to live and die for. And those things have not changed.

And so it is with each PECCA. People collaborate as much they can. But they also compete, because both competition and collaboration are in the DNA of the world.

The science of cognition knows this.[71] Tells us that perhaps "evidence is useless in settling controversial social questions".[72]

Let's stay with that, look at it. Repeat it.

"Evidence is useless in settling controversial social questions."

Once more, to slightly lessen the chance that we'll fail to notice, or forget, or be in denial, or rationalise because it makes us feel better:

"Evidence is useless in settling controversial social questions."[73]

We can talk all we want. Facilitate. Mediate. Persuade, bring around a great many, or be persuaded, come around ourselves. Reach something approaching consensus. But there is a guard at the gate of perception, armed to the teeth. And beyond him, in the passages, a censorship official. And beyond him in the atrium, there is a very determined memory amender/eraser.

Nobody can be made to believe what she or he doesn't want to believe, and there is a natural limit on the level of collaboration that can be achieved.[74]

Unless perception has a back door.

15

THE TRUTH IS HEARD

It's as humid as before, perhaps more, and they dry very quickly once they're back on the deck.

The yacht is almost dark now, and the night silent. Almost holy. A vessel with six souls in a vast expanse of water, delivered somewhat to providence.

"I can go and fetch a torch," says Juan-Franco, "or you can tell me what it says in those pages on the mast."

Julia sighs as she makes herself comfortable again. Having looked at the yacht from the water, everything on it seems smaller now. And the sealed document on the mast seems tiny. Quixotic.

"We have to know what's coming," he encourages her. "How to deal with it, what our options are."

She knows he's right. That conveying it is one thing, and the knowledge itself is quite another. That the fact that nobody's bothered to read it doesn't make it less important. Quite the contrary.

"I'll tell you. The summary of the summary."

"I have time. And my attention span is old-fashioned."

She smiles, and wonders if he'd notice in the dark. "You know most of it already, I'm sure," she starts, "but anyway. We're facing a very low pressure system that sometimes occurs in the summer months, close to the equator. In a low pressure system, air flows towards the centre – inwards, in other words. The closer you get to the centre or eye, the stronger the wind and the higher the swell. A big drop in the barometric pressure is one way to know that it's coming. But by then it's usually quite close."

She continues with added urgency, "Close to the eye and in a category five cyclone, wind speeds can be more than 150 knots and the swell up to 70 feet. The swell in the eye would be approaching from multiple directions because from everywhere outside the centre, everything is pulled towards the same point."

"You don't want to be there," Juan-Franco summarised dryly.

She agrees, repeating his words. She pauses, continues: "But you'd be lucky to get there. One extra sight-seeing trip before the end. Surfing a 70 foot swell in a yacht without pushing nose down and pitch-poling isn't that easy."

"So the swell we're having below us is aiming at the eye?"

"Exactly."

"And we're being carried in as we speak?"

"Yes. The sea's far too deep here to drop anchor, as you know. And so we're being carried, and the swell will grow and grow, and the speed at which we're going."

"But there's wind now, can't we not sail to the side of all this?" Juan-Franco asks.

She gets up so she can explain better, gesticulating. "I checked earlier...," she resumes, "the swell is going about northeast. It will move in a curve, changing to north, then northwest, then west, close to the eye. This means the eye is approximately north of us and we are in the so-called trailing quadrant. It could have been worse: the cyclone isn't moving towards us. It's slightly to port of where we're going now. We could turn about 125 degrees to starboard and sail southeast. That would mean we're heading away from the eye. And if we're lucky and it doesn't change direction and we haven't left it too late, we could gradually head further south as the swell eases, and we'll miss it."

"And the reefs?" asks Juan-Franco, remembering what she said before.

"The island is northeast of here. So southeast should be open seas and happiness."

Juan-Franco rises, restless. She sees him itching to take action, but realising they need a full crew and consensus.

"How much time do we have?" he asks. "When is it too late?"

"While we're heading downwind with the swell, we're hastening our demise. The only thing we could do to make it worse would be to raise the sails and fly the spinnaker. We have two chances; we could sail or motor out of this; or we could drop a para-anchor, try to stay stationary and hope the eye continues to move away. Both possibilities require that we turn the bow into the wind. We'll have to broach. If the swell is too high by the time we attempt this, we might capsize. So it's too late for us when it's too late to broach."

He looks out over the water, puts his hands down to stabilize himself as he leans forward. As if to meet what's approaching. Even in the darkness, his silhouette is asymmetrical.

She waits in the way that a lawyer waits for a jury. Having made her case. Waiting to hear what another thinks of it.

"And we have a para-anchor?"

"We should. If we don't, we can improvise one."

Juan-Franco thinks this over. If they had it now and dropped it immediately, it would slow their movement and give them more time. But there is Sam.

"Sounds like we're going to have to sail out of this while we can," he concludes. She nods.

"It's still very doable. If we start first thing tomorrow, all of us."

He agrees, speaks as if to himself: "We need to talk to everyone early. Perhaps we should get some rest. Sounds like it might be a very long day."

Julia starts moving inside, towards her suite. She assumes erroneously that Juan-Franco will follow.

Sam heard them returning after their swim and has been standing motionless ever since in the shadows, listening. He turns, gives two large, quiet steps and quickly gets back into bed, on his side.

What imaginations.

Early in the morning then. He'll just have to beat them to it.

16

THE DUAL DNA OF THE WORLD

Julia left the yacht for a while with Juan-Franco. They swam.

Returning, it appeared different. She was looking in, rather than looking out. There was another perspective.

She bobbed up and down. For a moment the swell was just a swell, not the precursor of anything. And it felt as if she was being rocked to a lullaby. Closing her eyes, she imagined the stars as they must be above the cloud cover, bright and reassuring in the way they're ordered – into the same constellations as ever.

She looked at the yacht, where some lights had been extinguished, and the perspective of not being on the vessel created the illusion that she wasn't part of it. Not its past, nor its present, nor its future. And that away from the yacht, there was a oneness, a benevolent harmony, gently enveloping her.

Like the yacht is to Julia for the time being, the earth is our home. For millennia, people have fantasized about leaving it. Looking in, rather than looking out. Gaining another perspective. And then space travel began and we saw breath-taking photographs of the blue planet in all its splendour.

So let's imagine the earth as a yacht. Outer space as the sea. And let's swim away with Juan-Franco to look in. Try to see from the outside what things are like, what their nature is, what the DNA of our world is. Because on this planet is a yacht with six people whose fate will be determined in the next 24 hours. By their collective ability to pursue the best course. And this alone makes it important to understand what the planet allows and what not. To understand what is normal, what is possible, what is unlikely, and what is impossible.

This is even more important, more crucial, because beyond the six people on the yacht are another seven billion. For some it is early morning, and they are rushing to work. Feeding their kids and getting them to school. Wondering about the health of their parents. Worrying about a friend who's getting divorced and another who's been left bereaved.

For others, it's late afternoon, and the sun is setting over a busy day. Or even a frantic one. Esthia is walking home through the park with a knot in her stomach. She struggles to breathe, and wonders who she can tell about it. But there is no-one she can confide in, nobody who knows and fully understands. Except Juan-Franco himself.

Millions of others are on their way home. Happy homes and sad homes, palatial villas and rain-drenched shanties, homes of abuse and homes of love. Crowded homes and lonely homes – like Esthia's will be for her, that night.

We're swimming away with Juan-Franco, into outer space.

"Okay,"you say, 'but give me a rope and a lifebuoy to hang onto. I don't want to drift into outer space. I'm scared of currents. This is my only home, my only hope, the only place where I can survive. Tie me to it while I swim away, so I can marvel without being afraid."

And together we swim a bit further away, bob up and down on the swell. The blue planet is about the size of a beach ball now. Smaller than us. We swim right around it, and notice darkness on the one side, and remember that in this half that includes the yacht, some are sleeping in loving arms, some are scouring through garbage in dark alleys for remnants of food, some are praying for salvation, some are getting ready with shivering hands to mainline heroin.

Juan-Franco has been silent. "It's not just the people," he says simply. He's eerily comfortable, hanging onto the buoy with his good arm. As if he's swum here many times before.

"Tell me," you say, closing your eyes, and allowing the swell to gently rock you, as if to a lullaby. Then opening your eyes again and seeing the stars, much brighter than ever before, but in the same constellations as always.

"You know already," Juan-Franco says, "but I'll remind you. I will tell you three stories. They are all true, but different. There is the story of the whole,[75] and the story of the parts.[76] The first is only imaginable from where we are now. Swimming in outer space. Hanging onto a lifebuoy tied to a very long rope, tied to our world – so the current won't wash us away. Seeing our planet as small as a beach ball. Smaller than ourselves.

The second story, the story of the parts, is easy to imagine, because we live it every day.

We'll start with the happy story of the whole, then tell the wild story of the parts. And then there is a third story that will scare you. And at the end of the third story, I will surprise you, and scare you even more, and show you that you've loved deeply without really knowing it."

Juan-Franco is momentarily liberated by the fact that this is a thought experiment only. He has a lightness, a playfulness that Julia hasn't seen and that Esthia has nearly forgotten.

"Behold!" he says, pointing with both hands at the beach ball, small but so far away that our rope soon becomes invisible, a fraction of the way there. "Ooops!" he jokes, grabbing at the buoy that you're clasping ever so tightly, taking no chances. "You almost lost me there..."

His smile fades then, the playfulness wanes. He's trying to keep it light, but he carries the truth he's about to tell, carries it like Sisyphus carries the rock, or Atlas the earth.

"You know about all the people. Seven and a half billion. Tall and short, thin and fat, dark and light. Two hundred countries, and many more languages.

Then there are our close friends; dogs, cats, horses, cows, pigs, goats, sheep. And our family; chimps, gorillas, baboons, monkeys, orangutans.

And so it goes on, from the ones we live with, to ones we know, to the ones we haven't even discovered. Maybe ten million species with billions and billions and billions of members across the species. And all of them sharing one vessel. A yacht in the ocean of space. One planet."

He stops and looks at it, lost for further words. Looks as if he needs a cigarette. Then looks some more.

"Now let's park this very rich planet, and look at a poor one." He turns around to point in another direction, kicking to swivel the lifebuoy, and a grey mass becomes visible. "We'll call it Daisyworld and give it only two species on its entire surface. Both flowers; white daisies and black daisies.[77] And let's assume we have a sun with remote temperature controls that we can adjust exactly the way we want to. Warmer, colder, more or less. In any sequence we want."

Juan-Franco pulls up a small console previously hidden on the lifebuoy. It has a simple, single dial, circular like an oil heater. For demonstration, he turns it clockwise and a very bright sun illuminates Daisyworld. Immediately he turns it down again, and the light fades and disappears.

"Okay, so a very low setting," he says, fiddling with the dial. Sparse sunlight falls on Daisyworld. "All we see is grey, for while the seeds of black and white daisies lie below the soil, nothing is as yet growing, and there is no life at all.

Remember now that a black car is warmer than a white car. Just in the same way, white daisies reflect the energy received from this sun of ours, while black daisies would absorb the energy. Thus, white daisies would cool the planet while black daisies would warm it. Just like a black car warms the people inside and a white one doesn't.

If the proportion of white daisies on the planet increased, the planet would cool, while relatively more black daisies would make the planet warmer.

So far we've been stingy with our sunshine. This planet is quite cold. The sun's rays are too weak and Daisyworld is too cold to support life. It stays just the way it is. Grey and barren. So let's turn the dial, shine a little light, live and let live, buy a little sunscreen."

He adjusts the dial and the sunlight on Daisyworld increases. Juan-Franco continues, somewhat absent-mindedly as he concentrates on regulating the heat on the experimental planet. "I can adjust time also," he grins, and points at a secondary control on the console, this one linear. He pushes it to the right.

"That's about two months," he explains, "Look!"

As we watch, Daisyworld is transformed. With just enough warmth to sustain black daisies with their superior ability to retain heat, the planet changes colour from grey to black as black daisies colonize it. The seed of some white daisies germinate, but then the white daisies die because they fail to retain heat. Their place is taken by black daisies.

Juan-Franco turns up the heat another notch, and slowly pushes time forward by another two months. "Warmer still," he explains. "It's becoming white daisy season." And sure enough, as the planet overheats a little, white daisies take over the entire surface because they can reflect the excess heat. The black daisies, unable to do so, die off. Their place is taken by white daisies.

"This is the story of the whole," Juan-Franco continues, putting time further forward and then putting down the console and turning to look directly at you. "So far it isn't that remarkable. All it has done is prove that black daisies like the cold better than white daisies, and the other way round. But watch what happens in time, without further interference from us..."

And sure enough, white starts to become grey. "The white daisies are now so prolific that they are cooling the planet," he explains, his voice getting excited, as if he'd never tire of the part he was about to tell. "And back come the black daisies. It has become inhabitable for them again, despite their getting more heat from our sun. And if we turned it down too low, the opposite would happen. The black daisies would become more prevalent and save the white daisies by cooling the planet through absorbing more heat. So despite slightly drunken or irresponsible or flippant people at the control of the console, the planet is able to survive. White daisies are life insurance for black daisies, and vice versa. The planet as a whole operates better the more species there are. The greater the number of species, the better they all tend to flourish. The planet as a whole, tends naturally to optimize conditions for life."

He turns us around again, back towards the blue planet. "Imagine now," he enthuses, "not two species of daisies, but ten million species of bacteria, protista, fungi, plants, animals, all interacting with non-living matter such as stones, in the most complex, unimaginable ways. Imagine the stability of that. The diversity. Always tending to make the whole more habitable, more friendly to life. New species. That is the beautiful story of the whole; in some mysterious way, it is all one.[78] Everything influences everything else. The whole survives and in the long run, it thrives."

He's obviously got to the second story now, the story of the parts, and he's pensive. He puts both arms on the buoy, hands together, rests his chin on it.

"You don't like this story as much?" you venture, guessing.

He nods in agreement. "The wildness I like," he says. "in some way. In my head, not so much in my gut. But it's true, so let's tell it. For the most part, the story of the whole is easier to comprehend when you're so far away that the planet shrinks to the size of a beach ball. What you see is harmony and a sort of higher order, planned or not. Black or white turning to grey, then more or less staying there. And it's easy to be magnanimous: life proceeds, life conquers, life diversifies.

But drill down, come closer. Closer and closer and closer. You end with a predator and its prey. A world red in tooth and claw. A desperate fight to the death, every single moment of every single day, over and over and over and over. Right now millions, maybe billions of members of all these species are drawing their last breath, trying in vain to fight off a stronger opponent. Becoming food.

In the story of the parts there is no magnanimity, no mercy. Just a desperate and doomed struggle for survival. A keeping together of a group of molecules referred to as 'me' for as long as possible. And then a recycling."

***

He's become quiet. Either pensive or forgetful, for he said he'd speak about other things, too. He starts to paddle back, towards the blue planet. Space swimming is far quicker than the sea equivalent, and we make rapid progress. Already the planet is far bigger than us.

"You spoke about a third story, and about scaring me, and about love," you point out, tapping him on the shoulder so he'll stop paddling, stop going back so fast. This thought experiment isn't over.

"Imagine a cross between Daisyworld and the blue planet," he begins reluctantly. "There are 10 million species, but the temperature is determined by the daisies. And so all the other species are dependent on them for survival. People too have to hope and pray that the daisies will keep the world not too warm, not too cold, or all will perish. And imagine the black and the white daisies finding that console we had just now, and starting to fight over it. The black daisies turn up heat to the max, hoping that their kind will prosper and the white daisies will die out.

Just then the last white daisies capture the console and turn the heat way down so only white daisies can survive and all the black daisies will perish. And so on. And on, and on. And in their millions the other species, including people, start to die.

Then too hot, then too cold. Caught in a war of daisies with no say at all."

***

You respect the agitated look on Juan-Franco's face, but while he rests and reflects and recovers his good mood, you paddle backwards. "I'd like to see it from really far one more time," you explain, realising a return is near. He relents, and soon the connecting rope is taut and the earth really small.

"There is collaboration in the DNA of our world," he concludes. "And there is competition. Always both. Infuriatingly, always both, even though we don't like it. Even though we wish it could have been the one or the other."

Then he lets go of the buoy completely, swims in a little circle, returns.

"No currents," he claims.

You look at the earth, small and distant, blue and brilliant, and at the black vastness of space, empty. Completely empty. Without further explanation Juan-Franco loosens your grip on the lifebuoy and pulls you a few feet away.

He's just explained to you that it's safe, but you still get a fright. You still want to stretch and grab the buoy and hang on.

Exercising all your willpower, you refrain from doing so and tread the waters of space, noticing that the distance to the buoy is remaining about the same.

"Tell me now if it's worth conserving. Is it worth saving?" he asks. "It could slip away any moment. Or you could. Tell me if you love it."

Then he swims back to the buoy, extends a hand, pulls you closer so you're hanging on again. You bob up and down. Looking at it like he does, it's far away and a whole, not parts. You look at the earth and feel the waters of space and see the stars and their constellations and you feel a oneness, a benevolent harmony, gently enveloping you.

A while later you notice that he's restless. "We're going back, aren't we?" He nods. You think of the competition and the collaboration, of the whole and the parts, of the yacht he'll soon return to.

"And so the world has its DNA the way it is, and you might be friends with many, but not with everyone?"

Juan-Franco nods.

"Not with Sam," you assert, and see him raising his eyebrows. Perhaps in surprise, perhaps in grudging acceptance of an existential oddity.

17

The faith of Lorenzo's grandfather

It's very dark now, with everyone in their beds and bunks. Even Juan-Franco, who came in last from the deck – hours after everyone else.

The wind has been picking up steadily and whistles now through the rigging. They should be getting up, setting sail, getting out while they can.

The swell is also growing. Growing. But it keeps running near perfectly from the stern to the bow so that there is very little roll. Before Julia fell into a fitful asleep she thought about this again. It couldn't be coincidence, and they were being pulled into the eye. Had to be. Wind and swell from the same direction behind them, keeping the yacht roughly aligned. But the day has been so quiet that there seemed to be no point in being ready the moment the wind got up. And now they're all still reacting as if to a perfect, quiet day. Not even anchored, for the water is too deep. Nobody at the radio, still outside reception. Drifting with the current. Just drifting, wherever it might be going, for better or worse. For worse.

Lorenzo is woken, the clear image of a running tap in his mind. He wonders if he's been dreaming or whether it is a vision. Closing his eyes again, he sees it clearly, the tap running. His grandfather standing over it, facing away, towards the livestock. Watching the precious liquid running into the trough, and the emaciated animals lining up in the dust to drink. The trough never filling up with the great thirst of the many animals, the tap just staying open, the water running. And his grandfather watching, waiting for the police or the sheriff to come and stop him.

Why now? The memory has haunted him for decades, but he doesn't understand why it's returning to him just at this moment. He is worried yes, but worried about other things; Sam, Julia. The radio and what the weather will do. And Angela, who is always looking at him and holding his hand when she's scared.

He feels the even pitch of the vessel, perhaps every seven seconds, rising on the swell. Stern, then bow, stern, then bow. It would normally make him sleepy, but somehow his grandfather is with him tonight, and causes him to be alert. Knowing that the next day he'd wonder about it. Probably think it was a dream so real that it felt as if he'd woken up.

Lorenzo never knew his father. But it didn't matter, for his maternal grandfather was there, on the farm, with endless patience and wisdom. The cattle were there, grazing, in occasional need of a dip, but ready to be milked every day. And above all was the order his grandfather taught him about; God in his heaven, man on earth taking care of creation and the animals. Simple and sure. Everything having its place and everything ultimately being alright. Lorenzo would look to his mother and grandfather for all he needed and they would provide. Collectively they would look to God, and He would provide.

And so it went, through better years and harder ones. Bountiful and meagre. For that is the nature of things, they move in cycles. So his grandfather taught him. Fat years and thin years – sometimes up to seven in a row. But always a return to normality. Always, in the end, God taking charge and stepping in and bringing it back to normal. For not even a swallow dies without it being ordained.

He was 14 when the drought came. That year was a hard one. The next year it didn't rain much either. Once, nothing at all for two months in the middle of summer. The animals started losing weight. And then the rain stopped completely. For a whole year they woke each day looking up at the sky. Feeling the air, for the raised humidity and eerie quiet that said it was coming. Rain, big rain, life giving and liberating and cleansing. And for a whole year it never came. It would threaten yes, and promise, and sometimes fall just on the other side of the mountains, not twenty miles away.

The neighbours started culling their cattle when the fodder ran out and had to be trucked in, and the city council promulgated ever stricter water conservation measures for municipal water. While the wells lasted that had no effect on Lorenzo's grandfather and his cattle, but then that too, ran dry, and all that remained was the municipal water, and nobody was allowed to use more than five gallons a day. When people failed to comply, the price was raised very high and a sliding scale was introduced; those who contravened were severely punished. At the start of the next year they even made it a criminal offence to use more, and threatened possible imprisonment.

Each day the sun rose in a perfectly blue sky and each day it set all orange and red, coloured by the dust lying on the plains all the way to the horizon. And even at dusk the red and orange image would oscillate as the heat lay on the land like an unwelcome blanket.

His grandfather just carried on as always, sure that the implicit order of things would be restored. He refused to discuss culling the cattle. In the evenings after sunset, he'd walk out and talk to them. Then take what money he had and the truck and drive the next day to buy more fodder.

All his cattle had names, and Jerry died first. One morning, in the dust. That's when his father walked to the tap and opened it and let the water run into the trough, watching the cattle line up to drink and drink.

The municipality must have had a water meter, for the police arrived before the sheriff and took Lorenzo's grandfather away. The sheriff only came the next day with a gun and the shots rang for a long time as he walked, shooting the cattle. Jenny was the last.

The next week Lorenzo and his mother moved to the city and he never saw his grandfather again until the funeral two years later.

Lorenzo studied commerce and became an expert in the financing of water infrastructure. He found a job at a development bank that does much work in Africa and the Middle East. He wasn't so sure any longer about God in his heaven and the fact that things always return to normal again. He never prayed again. Things change, sometimes.

And now, in the dark of this humid night, he saw his grandfather again, and the running tap, and the animals drinking. His grandfather watching them, waiting for the sheriff or the police. Lorenzo realises that the old man may have been facing away from him that day, but that he was smiling.

Why had he never realised that before? And why had he never even thought of that before?

But now, the tap, running – the vision or the dream. Why now? Is it time that is running out? Or is it a warning that something is about to stop being normal?

Closing his eyes again, Lorenzo starts to pray.

18

Bruce Willis and Green Die Hard

Sure enough, Julia is right. She's read the signs, applied the science. There is a Category 5 cyclone northwest of them. It's called Fifi, and it's already wreaked havoc wherever it's been. Its murderous aftermath is all over the media. On an island similar to the one they're heading for, in the destroyed villages, the misery is only just starting; water supply has been disrupted and cholera has broken out. Médecins Sans Frontières is flying in doctors and nurses, the international community is pledging assistance, an expatriate hip-hop artist is contemplating a return from his tax haven and running for president. Engineers Without Borders International is flying in all manner of technical personnel. They say in their media statement that there is a unique opportunity to completely re-sculpt the infrastructure to the benefit of the people. This is a euphemism that describes the fact that Fifi has put a small island back into the Stone Age.

What's left is practically nothing. Nothing. A few modern buildings built to German building standards in the last decade. Like the presidential home. Beyond that, there are tonnes and tonnes of rubble. Stuff once dug out of the ground and made into useful things like bricks, mortar, water pipes, corrugated iron plates. Now lying on top of the ground in a mangled heap, often on top of dead and maimed people.

Of a budding civilization, nothing is left except the memory of how civilization emerged the last time round. And perhaps the experience from Engineers Without Borders International – about how other people in other places did it.

The president is a man of some vision who has worked ceaselessly to uplift his island to the level of a developed country. He sits, smoking for the first time in years. He hasn't slept in days, and as he stares at nothing in particular from bloodshot eyes, he forgets about the ever-lengthening ash. The minister of finance fears that it'll drop on the Persian carpet in the president's office. He hands the president an ash tray that was always within easy reach. The hint is less than subtle, but the president takes it and the Persian carpet survives.

"What shall we do then?" asks the president, drawing deeply on the cigarette now that he has remembered about it.

"We must bury the dead and feed the living,"[79] says the minister of finance.

And utterly prosaic as it is, it is true. There is absolutely nothing to be done, other than to start all over again. Feed the living, bury the dead. The president thinks about his life's work, about the dream of uplifting his people, of building a modern, knowledge-based economy, of improving education and upgrading health-care. He knows all that is far beyond his lifetime now. Perhaps 50 years away, perhaps a hundred. He knows that in a year or two, when the severity becomes obvious to all, the best and brightest will start leaving the island and look elsewhere for opportunities that might perhaps manifest in their own lifetimes. They won't come back. The less educated will stay behind and frustration will grow when things don't really get better. Politics will move, the right wing and left wing parties will get stronger, the centre will fold. The language groups on the island will organize their politics around language and culture, and there will be conflict – and perhaps even civil unrest.

"Let's feed the living then," says the president, looking at the minister of finance, whose demeanour has changed only slightly since before the storm.

Some are phlegmatic. Some are not driven by passion or vision. Some do not need beauty to survive. Truth and falsehood are enough for them. They are valuable, especially in times like these. The president hasn't quite finished his cigarette, but he forcefully extinguishes it, squashes it in the ash tray.

"And let's bury the dead."

***

On the yacht, things are a little different. There is a storm, yes, and it is close, and it is voraciously wanting to pull everything it can into its centre and devour it – but it could be dodged. There is a clear way to avoid the cyclone, and Julia has already spelt it out and tied the answer to the mast. They could even employ the engine until they're safe. What prevents the avoidance of the cyclone and its concomitant destruction is the absence of sufficient consensus on the way forward. This again is caused by lack of information, pre-existing prejudices, vested interest, and manipulation. Soon there will be other reasons added; propaganda and coercion. And steadily the situation migrates to the point of no return, steadily the yacht drifts with the current, to certain destruction.

On the yacht, a situation where the dead must be buried and the living fed is still avoidable. But not forever. Not for long. At some point soon, very soon, the point of no return will be reached, and the yacht will be in the same position as the island. Perhaps worse, for islands cannot broach, can't do death rolls, cannot capsize or hit reefs, can't sink. On an island, the eye of a cyclone is a place of complete quiet, surrounded by walls of screaming wind and rain. At sea, it is the place where tsunamis converge from every direction to wreck everything that may still be defying Poseidon's promise to Hades – to surrender all that sails in that part of the ocean.

And so, when everyone including Julia is already inside, Juan-Franco sits down and starts to improvise a sea anchor. There might be a para-anchor on board, yes, but there's no way to know. Everyone is sleeping. And Sam won't even want to entertain the idea that a para-anchor might need to be thrown from the bow to keep the yacht facing the waves, and keep it as stationary as possible. By the time it becomes essential, it might turn out that there is no para-anchor on board. Or there may be no time to look for it.

Juan-Franco looks for maximum drag and weight.

He doesn't have much; a plastic cover used to cover the deck in port, a length of rope, a small toothpick anchor for use on calm days in shallow water, a net of unspecified use – perhaps fishing. Working as quietly as he can and sometimes having to strain his weaker right side, he wraps the plastic cover and the small anchor in the netting, then binds the rope tightly around it, adding as many knots as he can, yet leaving sufficient rope to let the concoction get away from the yacht a bit, should it need to be put into service. He pulls the plastic cover out a bit from the bundle to add drag, and then has to stop. For anything more sophisticated he needs to go looking, disturbing all those sleeping ... and invite opposition. What he's made is very rudimentary – but probably just enough to keep the bow facing into the waves.

Where to stow it? It's somewhat amorphous but still large, and wouldn't fit in the windlass hold near the bow, where he wanted it. Nor in the seat locker below the cockpit. He has to settle for the aft hold, working it to the bottom of the other items, so it isn't so visible. It would take time to dislodge and take out, more time to tie it to something at the bow, more to cast it overboard. But that is what he could do.

And so the yacht drifts towards disaster, with only the flimsiest defence against the cyclone that has just destroyed an island.

Beyond the yacht there is our world, our fragile yet beautiful, blue planet. Drifting into irrevocable climate change. Moving with nobody at the rudder towards catastrophic biodiversity loss. Looking at the post-cyclone situation on the island, where far beyond our own lifetimes we are left bereft of all beauty in a diminished world able to support a fraction of the people it did before. Some say for a 100 000 years. Others say 20 million years.

That's a long time to go without beauty. Or as Leonard Cohen may have sung without necessarily referring to our civilization:

"...hey, that's no way to say goodbye."

Three times now Julia has come forward with her truth, and three times she's been rebuffed. And many times scientists have come forward with their truths about our world and where it's going, and as many times they have been rebuffed. Juan-Franco has started to make contingency plans to turn the yacht back the way it came from, so it can survive.

Likewise, some people have started to make plans to reverse the way our civilization is heading, so it can survive. They see Green Sisyphus for what it is – a fruitless and endless repetition of the same actions. Some like Green Gandhi better and run campaigns and stage protests. Others think the cyclone is really close, almost at the point of no return. They feel it is time for violent resistance, [80][81] and they don't care if the authorities label them as terrorists, or if their punishment is determined by this epitaph. Their extreme concern and sense of urgency lead to extreme opinions and potentially extreme actions.

Two co-authors say:

> The claim that we currently are witnessing an advanced ecological "crisis" upon which the claim for revolutionary struggle rests, means that there is an emergency situation in the ecology of the Earth as a whole that needs urgent attention. If we do not address ecological problems immediately and with radical measures that target causes not symptoms, severe, world-altering consequences will play out over the long-term period.[82]

Another prominent thinker adds:

> As a long-time grassroots environmental activist, and as a creature living in the thrashing endgame of civilization, I am intimately acquainted with the landscape of loss, and have grown accustomed to carrying the daily weight of despair... Do you believe that our culture will undergo a voluntary transformation to a sane and sustainable way of living? I didn't think you would ... I'm aiming at a far bigger and more profound target than the nearly twelve million cubic yards of cement that went into the Grand Coulee Dam. I want to examine the morality and feasibility of intentionally taking down not just dams, but all of civilization [83]

One with the mind-set of a militia who can be imagined to look and dress like Che Guevara warns:

> I was brought up to succeed. To get mine any way I can. Whatever it takes to get what you've got coming. And if you can't do it honest, do it dirty ...That's how they make us ... There is no time or place better than right here, and right now. There is nobody more capable or more justified than you and I ... There is no excuse for our passivity and we are the only ones to blame if we even fail to try ...The primary responsibility of the struggle isn't to be saintly and righteous, it's to win ... fu** a theory with no action. We got too many of those ... people need inspiration, something to believe in, something to live and die for.[84]

In the final analysis, the archetypal 'lone good guy' ready to provide a metaphor for violent resistance in the environmental fold is Bruce Willis, playing the tough and uncompromising cop John McClane in the *Die Hard* movies. Invariably he's caught alone somewhere with an evil gang of heavily armed terrorists/robbers – and not scared to take responsibility for the situation, even though he's hopelessly outnumbered and out-gunned and it really shouldn't be his responsibility. The law isn't there, except in the form of McClane, who with considerable violence and the inexorability of the Grim Reaper takes out the baddies until the last (and most evil one) perishes in hand-to-hand combat and the crisis is over. Throughout, the credo that emerges from McClane is something like:

> This isn't my job, but there is nobody else to see that justice is done. So I'll do it as I think best. I'm not letting the bad guys win.

Green Die Hard. Individuals, and groups taking law into their own hands. And in the case of our blue planet, if we wait long enough, countries. [85]

***

Juan-Franco was busy for a long time, working on the sea anchor. It was hard going, requiring muscular exertion, but very little in the way of deep thought. Like earlier in the evening when he was playing guitar, his thoughts were free to stray. He thought a lot about Gandhi's words. Would it come to violence in the end? In the circles he frequents, he's heard some extreme views on what needs to be done to save the collective.

He has to take stock of all he's heard, read, thought. He has to remember everything he knows so that when the time comes, his thoughts are clear about the options and he chooses one that is bold and effective – but not excessive.

The swell's up higher, and the wind's getting up even more. Tomorrow will be the day.

As his fingers work on the sea anchor, his thoughts wander. He stands behind Esthia once more, at an esplanade. They look out over the ocean. She's holding onto a rail, and his arms surround her, also resting on it. Her hair smells of shampoo and vaguely of perfume. His chin is on her head, and she leans back against his chest.

PART 3

Ugly Lies

19

Sam's equaliser

This is a tight spot, thinks Sam as he rolls over in his bunk and feels the swell pass underneath the hull – lifting the stern, then the bow. It appears there is some weather system about after all. Not a raging cyclone for sure, but perhaps a moderate low pressure system. And with Julia having put the fear of God into all the others, it might just be enough to make them do something really silly. Isn't hysteria just the cause of every major meltdown? Damn it. It's the Fed all over again, over-reacting. Seeing a molehill for a mountain, and making it far worse than it would have been with well-intentioned and totally inept interventions.

Keep calm and trust the guys who can see the wood for the trees.

A tight spot. But he's been in far worse. He always survives, that's what he does, that's why other speculators call him 'The Cork'. In every derivatives deal, if it's big, someone must die. That's the zero sum game. What the one wins, the other will lose. That's what makes it exciting. And that's why it has to be other guy dying. Always the other guy. It's like a cowboy movie bar fight: you can take hits, you can go down, but whatever happens, you've got to be first on the draw, first to the gun. Then, afterwards, you recover for the next time.

In derivatives, the gun isn't a gun. It might be as little as a chat or a golf game with someone knowing something. A small present for someone who's deciding something. A bit of extra research through a private intelligence agency. A bluff, running the market one way with a rumour and playing them the other way. But you need guns and bombs. Small guns, big guns, hand grenades, nukes. To use depending on who and what comes your way. To survive for the next time, for the other guy ain't taking prisoners either.

This is a tight spot, but not so different. Julia's report is like a crop estimation report from a government agency. Frozen orange juice, maize, soya beans, whatever. Once the players know what it is, they usually follow it. They look for the numbers and the level of confidence and then they buy or sell. The big money is made before they know and before they understand. Find out what's in there before. Leak a false report on what will be in there. Create doubt about the level on confidence. Cast some shadow of impropriety over the official writing it. Whisper about conflict of interest or better yet, front-running or insider trading. Any of these approaches could make billions in a day or two.

He had to get them away from anything sounding like 'cyclone' or 'danger'. Strong wind is great. It is opportunity, it is the gods smiling down, it's what sailors pray for. Especially considering they could run dead downwind for maybe fifty knots and only beat a little east towards the island near the end. They'd be there as planned, tomorrow night, two days to spare before the option lapses. Having a seafood platter and Champagne at Sam's newly acquired six star resort. Being served obsequiously by self-effacing staff whispering to each other in the hallways that the New Boss is around tonight. There would be pretty young things amongst them, too. Pretty and easy.

Pretty easy.

Sam smiles at his own wordplay.

A swell passes under the hull, moving from stern to bow. About seven seconds later, another follows. But Sam is still thinking about getting to the island and then once on the island, getting action. He turns again to lie on his back, cups his hands behind his head.

Getting there in time will be pretty easy, too. He's reassured himself now, having thought it over. Just follow the rules. Arm yourself. Control what the plebs know. Interpret things for them so they don't have to interpret it for themselves. Tell them what they want to hear: This is safe, it's so exciting, this is exactly what Julia predicted, just much less of a big deal. Category two at most. Sorry to disappoint folks, no more excitement than that, but let's get there quickly now – dinner and drinks on the house. Spa treatments are on the house too. Anyone for golf in the morning? Looks like it might be a bit windy, but otherwise perfect, just like yesterday.

It's just Juan-Franco. There always has to be a joker in every pack. An idiot playing to touchy-feely bullshit about harmony, co-operation, inclusiveness, consultative decision-making. Whatever words you can find in whatever bullshit management bestseller about the eight and a half steps of success of all pathetic CEOs making money for other people. Blood-sucking shareholders. That is the one danger, the flank that's hardest to cover.

Arm yourself.

Sam remembers that he brought the 38 Ruger revolver in a Barsony shoulder holster. Just last week he was at the shooting range, practising with it. He'll strap it on below his shirt and say nothing. They won't notice and even if they do, it's for them to say something. Which they won't, because it's him paying for the charter. They are the guests, all here at his pleasure. And the captain is no longer with them. The one who at sea has the power over life and death. If he has to draw the gun to get them pulling the right way, his lawyers can argue that he assumed the vacant position of c aptain and was entitled to do so. If anyone lays a charge. And they can argue about being in international waters and which legal system applies, if any, and try to extradite him to wherever that may be and before the circus has got out of the starting blocks everyone will be exhausted and it will all just fizzle out. Like it always does.

The equaliser in the holster then. And an early start, before Julia and Juan-Franco can get going.

20

ARTISTIC MURDER, RATIONAL SUICIDE

Socrates and Nero both divorced beauty from truth, and as their lives wound down, the consequences soon became obvious. Socrates was steadily engineering his own demise while ignoring the urging of his dreams that he should make music. Nero, criticized for his excesses and especially for the way in which he seemingly celebrated the Great Fire of Rome, turned to propaganda – the ugly lie. He found a scapegoat in the Christians, blamed the fire on them, murdered them in the most artistic way he could conceive, and for a while united the populace behind him. Until the worm turned.

Initially, all he felt was a giddy elation. Oh, how the vulgarity was utterly vanquished, cleansed by the flames! Troy disappearing, crumbling, imploding. Making way for glorious, glorious Athens.

There is so much crudity in the world. So much that is devilishly ugly and quintessentially boring. He should have known it would take days for the flames to devour it. Nero thought of himself as the instigator and shook his head in near disbelief at the thought that he could be the one chosen by fate and history and the gods to spark this new beginning, this true golden age where barbarism would yield to culture. Enjoyment and opulent living would be approached scientifically, would be designed into every building and the smallest detail of that building, and beyond buildings into the city itself.

He itched to share his triumph with someone. Only the six gladiators knew, and he'd already had them killed by other gladiators – so now his secret was his alone.

Seneca. He thought of his mentor, who'd led him when he became emperor as a teenager. The wise one who taught him about politics and about how to survive. He would have liked to tell Seneca what he'd just pulled off, so Seneca could understand what an excellent pupil he's been. But Seneca was in retirement, put there by Nero himself. When the time came to govern without anyone holding him back, he'd decided to spare the old man, and only accused him of corruption. Seneca got the hint. Understood that his charge could be taught no more and had to ascend alone to that higher level reserved for the very exceptional. And so Seneca quietly got out of the way, retired to his villa and spent his days writing.

Then Nero thought of Poppaea, his beautiful wife, whom he'd married two years earlier after a lengthy affair. She was the one to fully grasp his talents and glory. To know what really happens in the empire and what it takes to pull off the things that he, Nero, pulls off daily. But this ... this was truly unbelievable and if he told her, she would be itching to share it with someone – like he was itching. It's too big, too magnificent. He can never tell her.

And then there is Petronius, the writer. The Athenian amongst Spartans. The one who understands good taste and good living better than anyone else. The Olympian amongst voluptuaries. He too could not be told, much as Nero would want to. For an errant word one of the many times when he is intoxicated would give it all away.

But Petronius could be involved in every fine detail of what was coming. How to design the Golden House in the middle of the city, where the rubble now smouldered and citizens poked around for possessions and signs of life. The gardens and the vineyards. The flocks. And there had to be gold everywhere, and precious stones, and frescoes, and many, many rooms all designed for pleasure and debauchery.

More than anything, there needed to be a colossus, a massive statue of himself. Nero, the one who brings into the world this new era of gold, and precious stones, and refinement in every minute detail. Who creates the tapestry of this new sophistication and makes the heavenly music to go with it. Even the events, the orgies, would be pieces of art. Transient – but art incarnate nevertheless.

There were such statues in the ancient world. In Egypt, standing still, the Colossi of Memnon, two depictions of rulers at the River Nile, reaching 60 feet into the sky.

Greece surpassed this, as she surpassed Egyptian civilization. The Colossus of Rhodes had been almost a hundred feet high before it was destroyed by an earthquake three centuries earlier.

The Colossus of Nero would be the big one. The Big One. A hundred and twenty feet at the least. To commemorate Rome and its civilization surpassing Athens and anything that had ever been before. But it would have to be sculpted by a Greek, by a remaining custodian of that arcane knowledge of large statues and fine art.

Petronius would understand, thought Nero. Would walk every step of the way with him, would share his excitement, would see his genius.

***

Nero's concern for truth was diminishing from very little to nothing. Other than to understand if someone was plotting to kill him, he didn't care about truth or falsehood at all. Beyond his own survival it was irrelevant to him. Truth mattered only when something arose that might interrupt his quest for beauty. The more unusual any pursuit, the more skilled a performer, the more impossible a feat, the more dangerous, the more ground-breaking, the more appalling an occurrence – the more beautiful it was. Especially if he was the performer, the charioteer, the hero.

And because outside his own survival there was no truth and falsehood, there was no right and wrong. There was exhilaration and boredom – that was all.

But others cared. The people who'd lost homes and loved ones. They wanted to know how all this happened. And they heard stories that Nero had been observed singing *The Sacking of Troy* while the fire raged. They wanted to know if this was true. Wanted to know if it was true that gladiators had been seen setting fire to the city. Pretending to be drunk. And the people wanted to know if these things were True because if they were, they were Wrong. Soon, Nero heard that there was discontent brewing, anger, even hatred.

Something needed to be done to move the focus. And thinking it over, most gloriously, he suddenly saw that it created the chance to weave a new, heroic narrative of Nero. The Emperor descended from his palace and helped the survivors to search and rescue survivors from the blaze. So brave was he that he even left his bodyguards behind. For these wretched, the salt of the earth, these Trojans in heart and soul and demeanour and spirit, ultimately made up the audience to which Nero the Artist was playing. And thus, their adoration and respect needed to be cultivated.

For days he sifted through the wreckage with them, patted them on the shoulder as they broke down about what was lost, wept.

When there were no more survivors to find, no more dead to extract and bury, no more possessions to be salvaged, Nero returned to his palace and threw open the doors to the homeless – paying for their keep with his own money. Another glorious tale in the ongoing, heroic and glorious life of the Emperor Nero.

He thought that would settle it. Indeed, his thoughts had moved already to the next chapter. He'd ordered the slaves to move in, in large teams, to start clearing the rubble. There were long discussions with Petronius about what should be done. How the new palace would look. What marble to use, and what stones, and who would paint the frescoes. The posture of the sculpture, and where best to put it. What would be done first, and second, and third. The exact needs of the guests and how best to cater for these in the designs.

But then the rumours resurfaced.

"They're saying you were singing while the fire raged," said Poppaea, brows lifted to underline the gravity.

"But they saw me helping them for days and days!" he countered. "And I brought them into the palace afterwards and fed them. Have they got no gratitude, no sense of decorum?"

"If you were singing, then at best they think you didn't care," Poppaea pointed out. She was always more in touch with what the plebs were thinking. "And at worst they think maybe you started the fire."

Nero mulled this over in the days that followed. His mood was foul, and people stayed out of his way. Just now, in this critical moment when culture was to be reborn – this triviality, this utter pettiness. And that while there were always conspirators around, waiting for the slightest opportunity to encourage insurrection. Vermin like Piso, Lucan and Plautius Lateranus.

He'd thought that heroic action by himself would deflect these thoughts, but it wasn't working. The plebs were angry and they needed someone to blame. More; they needed someone to suffer as they thought they were suffering.

Nero thought about his enemies, about whom he could blame to get the attention off himself. It would keep the plebs happy and he could get rid of the enemy at the same time. It needed to be a person or group without political support, for otherwise punitive action might invite the very insurrection the emperor was trying to avoid. Something to unite the masses, not divide them. Someone to punish and hate for what has just befallen the plebs. Something the plebs would see as winning back the favour of the gods.

He paced up and down his room, the same room from which he'd watched the fire start and spread. Up and down. It was early August, hot and sultry, and he hated heat, for even as an Emperor there was so little he could do to escape it. Again he frowned. Swore. Kicked a chair, then swore at it for hurting his foot.

And then it hit him – the Christians. They were an abomination, refusing to believe in the Roman gods. Denying their existence and talking about one, true God. This tiny sect from Jerusalem, but spreading, growing. They were even saying that their God is all they can obey – not even the emperor mattered. They were the perfect target – all the plebs would agree with them being punished. It would be divine justice, given their refusal to acknowledge the gods. A fire started by the Christians would be exactly what the plebs wanted to hear.

From that moment, matters developed quickly. He had prominent Christians arrested within hours, and tortured immediately. They confessed – perhaps to being Christians, perhaps to setting the fire. It didn't matter. Word spread. "Did you hear? It was the Christians!"

"It was the Christians."

"My brother is a centurion, he says they arrested the Christians for starting the fire – and they confessed."

That took care of shifting the blame. What was left was feeding the beast of the plebs' anger.

In his usual fashion, Nero converged on beauty. On the captivating, the revolting, the novel. Anything that wasn't commonplace. Moreover, there was a chance to warn conspirators and rumour-mongers about what could happen to them. About the kind of punishment Nero could mete out to anyone suggesting that the emperor had been involved in the fire. So it had to be severe punishment in a novel way. Initially, he had the Christians crucified. Petronius, who liked satire, said this would be a good thing to do because the first leader of the Christians had been dispensed with in this manner under Tiberius three decades earlier. But Nero thought he could do better. So he covered them with animal skins and threw them to the dogs to be devoured in a frenzy.

But still he thought he could do better. The theme was fire – how could he kill them in a way more suitable to the artistic theme? Fire. So he burnt them alive in his garden after sunset, to serve as illumination.

The plebs were pleased at the punishment of those they despised. Shocked at the idea that similar fate might ever befall them for whatever reason. And their anger dissipated as the orgy of violence and torture went on, and on, and on.

The rumours died down and would not rise again until the history books were written with Nero departed. Then, they gained life again. There were questions: "If the Christians had been tortured into confessions, were they really guilty?"

"Why was Nero seen singing, and why, of all songs, 'The sacking of Troy'?"

Nero's atrocities would catch up with him at a villa in Antium less than four years later. He would suffer fear as intense as that of the people he'd sacrificed for starting the fire. And he would perish with a whimper.

But his ugly lie about the Great Fire of Rome would serve him well. It would be for other things that he would be hunted down. Not for artistic murder.

In Athens almost five centuries earlier, Socrates was planning another kind of death. Not of others, but his own. Not murder but suicide. Rational suicide – calmly having looked at everything. Making sure that it leads to victory in the trial he was conducting, the Trial of Human Nature.

It would take years to implement, but after that day in the Boule when as the chair he defied the mob and stuck with the Constitution, he knew they'd be coming for him – to drag him down to their level or to get rid of him. Exile him, or if necessary, kill him.

There was a final confrontation looming between himself and the capricious ones, who said "yea" today and "nay" tomorrow, with the only consistency their hatred for anyone who reminded them of their own fickleness and lack of character. And remind them he did, every chance he got, everywhere he ran into them. It was "I told you so," relentlessly. "I'm wiser than you". "I'm more virtuous than you". And, most obviously, between the lines, "I am more intelligent than you."

His students – teenagers and adolescents – couldn't get enough of it. They were at the rebellious stage of their lives, and their teacher mercilessly mocked and attacked the figures of authority they were rebelling against. Parents, senators, merchants. Socrates had the pride and dignity of the establishment hung, drawn and quartered in the squares and on the streets.

They wouldn't take it lying down. They'd be coming.

Again, the dream came – about making and cultivating music. The injunction to add beauty to truth. If it was from Apollo or Dionysus, could he ignore it? He was approaching seventy, it was a good time to obey the gods. Put perhaps it wasn't from the gods.

It wasn't from the gods – so he decided.

And so he didn't make music, didn't cultivate music. Still avoided the comic poets like the plague and went to the theatre very seldom.

With three sons now, his life with Xanthippe wasn't getting any better. She often accused him of not looking after his family. She wanted to know why he couldn't charge to teach, like the Sophists did, so at least they could live a little more comfortably. He was happy to reason with any man, but he'd long since stopped debating anything with his wife.

He felt tired sometimes, tired of a life of opposition. In the daytime, with the vacuous of Athens, there were fights be picked and enjoyed. But in the mornings and evenings, there were fights with his wife that she picked and that he didn't have the energy for.

Death, so he told his students, might just be the greatest of all blessings. There were many enemies in the city hoping to oblige him.

They had their first opportunity two years later, when the Thirty Tyrants grabbed power and democracy was again suspended. Remembering how Socrates had washed his hands of the execution of the eight generals, they immediately found a way to make him complicit in the worst of crimes, so that the verbose gadfly could be silenced for good.

Leon of Salamis was a perfectly innocent man having done nobody any wrong. "Go and fetch him and bring him to us to be executed," the tyrants ordered Socrates and four others. It wasn't a bad ploy. Socrates would be sacrificing all he believed in, and all his legacy. If not he'd be risking his own life. The decision was easy: Socrates disagreed and refused. The ball was back in their court – they'd have to kill him unjustly to regain the initiative. Before the drama could develop, the tyrants were overthrown, and Socrates emerged with his reputation further enhanced. He was the conscience of Athens – a highly irritating one with an utter lack of diplomacy.

The years rolled on and he became ever more confrontational. Then his friend Chaerephon got the idea to approach the Oracle of Delphi and ask her if anyone was wiser than Socrates. "No," the prophetess said emphatically, "Nobody is wiser than Socrates." This was fuel on the fire. Word spread. Armed now with divine confirmation of his supremacy, Socrates even went as far as praising Sparta, the enemy.

In this ancient form of Russian roulette, a plan was forming. He thought about the emerging pattern and where it would inevitably lead to. He was a threat to those who wanted to circumvent the law and the Constitution, and they'd keep coming until they got him. Until they'd exiled him or killed him – so they could have carte blanche. He was old and didn't feel like starting a new life as an exile. And he was tired, but wanted to win. Wanted definitively to demonstrate to all how he was right, and they wrong. He virtuous and they, evil.

What he needed to do, when they came, was to elevate the matter, to raise the stakes as high as possible, to allow only two outcomes: victory or death. And if he was strategic, he could manipulate it so that even in death, he would win. Which would leave only one outcome.

Victory.

So he thought. But he underestimated his legacy, and never thought his endgame might influence global affairs more than 2000 years later – in a way not even vaguely resembling victory.

***

In the interim, an amnesty had been declared and they could not accuse him of political crimes. It would have to be something else. For four more hot and dry summers he fired up the youth and aggravated the vain. And through the cooler autumns, when the vineyards turned brown and yellow, and then through the sometimes rainy and grey and chilly winters. He gave them no respite.

Ultimately, searching through so many of his words and utterances for the best way to get at him, when they charged him, it was with subverting the youth and not believing in the gods of Athens.

Suddenly standing in a court of law for the first time, with his life in the balance, Socrates felt a sense of unreality. For so long he'd known it was coming, for so long he'd prepared and provoked to create this final showdown. And suddenly it was totally upon him. All the faces of those who would be voting on his guilt or innocence. And one chance to validate a lifetime.

Reason and Virtue were on trial, and he was their advocate, and whether he survived or hopefully perished, Reason and Virtue had to win. Absolutely had to win. Indeed, if he perished, the chances would be so much greater. The execution of the innocent Socrates would be the glorious victory of Reason. In time, who wouldn't know and understand the truth? In twenty years, when his students were the most influential people in the city, who wouldn't remember Socrates as just and virtuous and conclude that a grave miscarriage of justice had occurred? And from there, it would be a short step to understand that the snap decisions of the masses spurred by the emotions and passions are always wrong. That any opinion held jointly by a huddle of drunken soldiers is wrong, without fail.

Meletus was the ring leader. The charges were very thin, but they would have to do. The complainants knew where they were going, knew there were many who supported them, and the charges would just have to stick.

It was well before noon when Socrates found himself commencing his defence. He started by making sure he'd get his full opportunity: "I am old and have never been charged with anything before," he started, his voice steady, "and so I beg you, even if I sound like a beginner in court, do not interrupt me and to let me finish what I want to say."

It would make no difference to his fate whether they listened to him or not – it would be exile or death. If he constructed his speech correctly, death. But to the innocence of Reason, it would make all the difference. He had to have a chance to say all that there was to be said, so that they could hear and remember, and so that his supporters in the audience could later write it down. Plato perhaps, or Xenophon.

With everything in place, he started speaking. Between sentences, as he stopped for breath, he remembered the times he went to war, and the things that happened. In daytime, the waiting, then suddenly the battles, the shouts and cries, the bleeding and death, the one side victorious and the other fleeing. At night, the huddle of laughing, loud soldiers. The inanity of their lives, and yet the scorn they had for him. The dream about making music. He thought of Xanthippe, her dark and dissatisfied eyes, her scolding voice. In that moment, drawing breath to continue the most important speech of his life, Socrates had the incomprehensible sense that she cared for him.

He went on, many words, more than eight thousand just to explain his innocence. He told how the Oracle of Delphi pronounced him wiser than anyone, and how he tested this by talking to Athenians from all walks of life, and how he confirmed that it was true. He, Socrates, was in some way superior to other men.

He was accused. Did he turn, look the jury in the eye and say: "I have no intention to subvert the youth?"

No. He questioned Meletus, used clever rhetorical and logical ploys to immerse his statements in absurdity.

Did he say to the jury: "I believe in the gods of Athens?"

It would have been easy. He could recount his recurring dream about cultivating music, and the distress it caused him that the giods were ordaiomething so far from his nature. But no. Instead of such an affirmation, he took the statements against him apart, little by little, with the aim of leaving nothing standing in logic. And at that point, he said that no elaborate defence was necessary.

As he stopped again for breath, he reflected that the defence of Reason was going well, but reminded himself that he had to raise the stakes. Plant the seed of execution rather than exile, for if the miscarriage of justice wasn't great enough, Reason might lose.

He proceeded. The sun was past it highest, the night was on its way, inexorably.

"I have many enemies," he said stoically, "and if I am destroyed it will be through envy, which has been the death of many good men. I won't be the last."

He went on. Explained that he had no choice but to do what is right, despite the possibility of death. That God had commanded him. He didn't say which god. And he explained that he was the gadfly, the one who kept them on the right track, that they needed him and would suffer without him. He was their conscience, they should not kill him. His long, long sentences were finally winding down. He reminded them how he refused illegality as the chair of the Boulle. How he took no part in the execution of Leon of Salamis.

Concluding, he again said that he would not defend himself with tears. That it would be wrong. He said that justice needed to be objective and not arbitrary.

He was closing, he was about to say something which might exonerate him, but again he chose to place himself a step above the jury and audience.

"I do believe that there are gods," he finally said evasively, "and in a far higher sense than that in which any of my accusers believe in them. And to you and to God I commit my cause, to be determined by you as is best for you and me."[86]

The jury quickly convicted him. As he knew they would.

The ugly lies against him stood, intact.

Preparing for the sentencing, he wondered if he'd said enough to show that it was a mistake, so that Reason could be exonerated.

Just before he started his address on sentencing, he thought again of the dream. Perhaps it was from Apollo after all. There wasn't much time.

He would have to do something about music, soon.

21

ESTHIA

When Esthia got in from the park the second time she was still shivering, despite the stiff walk. What to do? She could muster no energy at all for the article on the bee colony collapse.

It was exactly the type of situation tailor made for contacting Juan-Franco – for that rare telephone call that she made once in a blue moon when she was really at wits' end and needed to talk to someone she trusted completely, both in their discretion and advice.

She'd call him at a time when he was likely to be alone and he'd recognise her number, so by the time he replied there would be warmth in his voice. "Imagine this..." he'd say half jokingly, knowing how uptight she always was at the start of these conversations. And she'd start to talk back, trying to keep it light, but less successfully than him. He'd let it go until he could hear she was grounded, or if she wasn't getting there, he'd say; "Stop!"

And he'd let five seconds pass, ten. "Slowly...," he'd say, "... you're going too fast. Come down from the head stuff."

And she'd start over, breathe, and talk much more slowly – and a semblance of the strength and balance she had that first day in the park would return. From there he'd let her run, like a rider does a trusted horse, with just the lightest of touches every now and again.

What would she say if she could phone him now? It's impossible to separate Juan-Franco as subject and listener. There was no way she could say; "The person closest to me is in great danger," because she'd be talking to him, the person, and he'd know it. And anyway, in their totally frank discussions of once or twice a year everything would migrate towards full disclosure of everything. And if she couldn't keep the listener and the loved one apart, she could as well start by telling him the many things she'd never told him over the years because she's never told them to herself. Always shelved them somewhere, deliberately misfiled them somewhere so they need not compete with the sanity and order of her life. The everyday life with its husband, its kids, its job, its normality, its mediocrity.

She couldn't call him, and if she wrote to him and he survived, then all of it would be out there, said, and once said might acquire a life of its own, and everything would be very upside down. And if she didn't write and he didn't survive then she would have missed a chance to write. In a way still talk to him one last time. A chance to be there for him in some little way when he needed it.

Writing would be the right thing to do only if he wasn't going to survive. If he survived, having written would have been good but having opened the Pandora's box not. Again a motorist hoots at someone in the street below, again she freezes and remembers Samos and the ferries and the mist.

Esthia realises that she's standing, that she must have been pacing up and down in front of the window. 11:30. Wow. How did that happen?

She could decide later about writing, perhaps she could do something first to make sure he survives, because then all this goes away.

Esthia calls the coastguard and holds on for a long time while they look for the right person for her to speak to.

"Madam, that's way out of our jurisdiction," the man says after she's explained about the yacht and the cyclone. "I'd have said talk to the coastguard locally, but the island group is small and to my knowledge has not set up such a facility so far. My guess is their police and perhaps their navy would undertake those functions."

She's about to ask if he has a number she could call, if there is someone he could suggest she speaks to. But then she remembers the devastation she saw on television and knows it would be an utter dead end. The army and navy, if intact, would have bigger fish to fry on the neighbouring island. Potential riots, starvation, disease. She thanks the man and hangs up.

***

She does some overdue admin work. It takes very little concentration and she feels up to it even while being distracted. Lunch comes, and she goes down to the cafeteria. She's missing something, surely. No coastguard, the army busy, the navy overwhelmed too. There's no embassy but there is a diplomatic mission – but also abandoned on the main island after the storm.

Who else could she call?

She thinks of the people on the yacht. Juan-Franco has no relatives or friends who can be of help. She knows there's a couple on board whom she hasn't met. And a single woman whom she's maybe seen once, if she isn't mistaking her for someone else. And Julia, who moves in her broader circle of scientific friends and whom she sees occasionally. Julia's friends might be able to tell her more about the cyclone, or about yachts, or about cyclones and yachts – but none of them owns a search and rescue company. Or a helicopter.

The cafeteria food is well below par – some Moroccan lamb stew with hardly any meat and some grossly overcooked vegetables. She has one more bite, then pushes it away and fetches desert in the hope it will be more appetising. She feels queasy – perhaps eating something might help. Tipsy tart and custard.

In the order of people as she likes them, her thoughts finally drift to Sam, the gezillionaire. If he'd been in the office, he might have been just the man to call: "Sam, the yacht Julia's on is right in the path of a cyclone, and there is no other way to rescue her. You have to charter a helicopter right now."

And he probably would have done it. Would have gone himself, chartered a jet to nearby, helicopter waiting, all flash and Ray-Bans and new money, working on a lovely tale for late night at the bar at the next World Economic Forum in Davos – but he would have gone.

Now he's on the boat, outside reception. He cannot be asked to go. Poor weather and absolutely no mobile or radio reception.

But Sam is rich, and deemed valuable, and has to be insured. That's when the really bright idea comes to Esthia. The insurance company.

***

She has Sam's PA on the line at a quarter past two.

"I'm sorry madam, I am not allowed to divulge Mr Preston's whereabouts, but he is on a business trip with some associates."

Esthia starts losing her temper but fights it down immediately, together with the 20 or so super-sarcastic put-downs flashing through her mind. Corporate bullshit never worked for her.

"Madam, thank you," she starts very positively, now having to repeat the introduction that fell on deaf ears, "I am an associate of Sam's myself. If you scroll down his contact list you might find me somewhere as I have sailed with him and his wife on occasion."

She repeats her name, hears silence on the other, feels that she is winning.

"The yacht Mr Preston and his associates are on is in the immediate path of a cyclone, or very near to it. There is no evacuation service to call on the islands."

"Yes." says the PA, all attention now, coming round. This is her boss, and her job.

"I called to establish if his insurance policy includes a search and rescue service in the case of natural disaster, injury or death. If he gets sick in a remote area, is he flown out?"

"Yes, he has such a policy."

"Well, please call them immediately," says Esthia. "Please report the full situation to them, and get the wheels rolling immediately. If you need any help, call me."

She leaves her number, says goodbye, hangs up. Shivers and gets goose bumps for a reason she initially doesn't understand.

The line is dead, the call over, and she looks at the phone that she senses has just given her something. Then puts it down.

It was power and conviction. The way she'd possessed it that first day in the rain with Juan- Franco, and the days before, and the months after. And almost never in between.

***

She walks home, through the park, the knot in her stomach. Where she veered off to go the pond in the morning, she keeps walking straight ahead. It's cold, her coat is over her arm. Her head and neck are somewhat straight, somewhat forward. She pulls them back, straightens her posture. The way the ballet teacher taught her.

It's dark by the time she gets home and as she enters, she hangs her coat by the door, in its usual place. The dog comes rushing, then it's kids and her husband, dinner and homework and dishes.

She's back at the laptop just before 10, in her study. The house is settling – dark and quiet creep in. She puts on music and turns up the heater. Should she write? She starts looking for the Pablo Neruda book. Somewhat surprisingly, she finds it, notes how new it looks, how un-thumbed.

She resists opening it, but puts it on the desk.

Write she must, she can't do otherwise. She starts, stops, erases, starts over, a bit slower. She comes down from the head stuff, and the second time it comes out better. And then it just flows, an account starting when she was a child, stopping on a ferry, and carrying on for another decade and a half.

She saves the file sometime after midnight, and password-protects it.

This was the compromise she's come up with. To write yet not to write. Going to her email, she writes a short note to Juan-Franco, saying:

> Thinking of you all day. Heavy weather, I hear. Wrote you an unusually long letter. Phone me for the password.
> x Esthia

She sends the email with the long letter attached, and as a text to his mobile, without the attachment. Either way, if and when conditions allow, he'd get it. And she could still decide to give him the password, or to claim it was sent in a very weak moment, and keep the password to herself.

22

Propaganda: the Seesaw's Second Prejudice

On the empty deck, the plastic folder with Julia's ugly truth still hangs on the mast, unread and unheard. Writing it down was her way of communicating it. Previously, she gave two oral communiqués. So far, her communication has failed – mostly due to Blackbeard's Ghost.

We had Tom and Sally on the seesaw. Tom being heavier but somehow staying suspended, with Sally rooted to the ground. Then Tom's mother lending him a hand to come down. Failing at first, ultimately hanging with all her weight on Tom's side, also suspended. This would be an impossibility of physics, but it was explained by the ghost. Invisible but really heavy, and present on Sally's side.

Blackbeard's Ghost is our human prejudice for good news. The invisible weight causing the scales of rationality to come down on the lighter side. Thus, if truth brings bad news and implies that it requires big change, it starts at a disadvantage. The listener doesn't want to know. Truth begs for corrective action, but the lie says, "Relax, everything will be alright." The truth just wouldn't speak for itself. Ignaz Semmelweiz died; beaten up, rejected, alone, forgotten. It's not doing Socrates many favours, either.

Julia and Juan-Franco are in their bunks, sleeping the light and fruitless sleep of people who have stopped thinking, but not stopped worrying. They're out of action but not resting. In the few hours after they rise in the morning, on its present course the yacht will pass the point of no return and will inevitably be sucked into the eye of the cyclone. There, the chances of survival are slim to non-existent. So in the morning, they would have only a few hours to persuade and align the rest of the crew. To turn the yacht in another direction, mostly facing the wind, and to beat to windward, to safety.

Is there anything else they have to contend with, in addition to Blackbeard's Ghost? Once people decide to listen in the willingness to make big change if need be, how can it still go wrong? Indeed, it can go wrong in another way. Someone could steal onto the deck and replace Julia's exposition on cyclones with another that states false facts. Junk Science or outright lies. That says, as one example, that the wind gets weaker the closer one gets to the eye.

To return to the seesaw, let's assume the fulcrum and base of the seesaw isn't static. It can move on a rail, like a train – closer to Sally. And let's assume we put it in motion. Slowly but surely the base of the seesaw moves closer to Sally, and further from Tom. All the time, the total length of the seesaw remains the same – the part from Tom to the base gets longer as the base moves, and the part from Sally to the base shrinks as the base moves on the rail.

What's happening? Tom's leverage is increasing, and Sally's is decreasing. And quite soon, even though Blackbeard's Ghost weights far more than Tom and his mother combined, the seesaw starts to tip. Aided by the leverage, Tom and his mother outweigh Sally and Blackbeard's Ghost.

Now, let's reverse the process and move the base of the seesaw the other way, towards Tom. Sally and Blackbeard are the beneficiaries of increased leverage, while Tom and his mother's leverage is decreasing. The seesaw is developing a second prejudice. Soon, even before the base reaches the centre point between the two ends of the seesaw (and because of Blackbeard's considerable weight) the seesaw tips back the other way. And now the base continues to move towards Tom and his mother, giving Blackbeard's side ever more leverage and rooting Sally ever firmer to the ground. By the time the base stops moving, it's far closer to Tom than to Sally – and two more women the size of Tom's mother would not be able suffice to tip the seesaw again.

Recall that on the other side, Blackbeard is still invisible, and only Sally can be seen. To someone looking quickly (who notices neither the Blackbeard's Ghost nor the fact that the seesaw is unbalanced) the spectacle would be quite bizarre: Three women and a boy on one side of a seesaw, and a little girl on the other, outweighing them.

Blackbeard's Ghost is the prejudice of the listener – how the message is received. Thus, if it's a 50-50 call, Blackbeard's Ghost will usually ensure the good news is heard and the bad news is drowned. It's all about a beautiful lie.

The unbalanced seesaw is prejudice in the way things are communicated in the message itself. Deliberate error message. Ugly lies. Crass and effective. Often used not to convey information, but to achieve persuasion.

It relates to that part of our nature that tends to compete, rather than the part that collaborates.

The movement in the base of the seesaw is Propaganda. It causes the leverage of one argument to increase while simultaneously causing the leverage of a contrary argument to decrease.

In the case of science, there are controls over how people speak. There has been a development from 'might is right' to 'right is right' and 'truth is king'. The father patiently listens to what the son is saying, asks questions, challenges gently. The son gets the feeling that he is only half right and goes away to think, then returns with more nuance and confidence. And ultimately the father smiles. Pats him on the shoulder and says: "You are right."

There is a scientific method. It asks the son, before he speaks, to eliminate all personal bias, all emotional involvement, all prior loyalty. It then asks him, if possible, to measure what he's talking about. Or to find a way to test it. A way to show it up if it were wrong. And then to tell everyone how he discovered what he knows, so someone else can repeat the same experiment and find the same answer. And the son is asked, finally, to make predictions that depend on the truth he has discovered. If the predictions come true, he knows it to be true, and all the world of 'right is right' agrees with him.

Science has found a way, when it concerns things that can be measured, to ensure that most of the time, speakers speak the truth. This is what Julia did when she wrote the truth about cyclones and tied the handwritten notes to the mast in a plastic sleeve; she rendered the information as accurately as possible. She tried to take all emotion out of it. She wrote down the cold, hard facts. They may not be heard because they are ugly, but they are true.

Propaganda is what Sam will do to obscure and twist what Julia has written down, in order to solicit a certain response from the people hearing it. A response that suits him. One that causes him to survive, and the other guy to die. Always the other guy.

***

The threat to Julia and Juan-Franco is a cyclone. More particularly, whether it is really there, how strong it is, whether it can be avoided, when the decision needs to be made, how to convince the others.

On a global scale, the threat can be climate change or biodiversity loss. Both environmental challenges that require collective action to be resolved.

Collective action on a yacht or in a village is one thing, in the global village quite another.

Imagine we all live in a former fishing village that is home to only 150 people. It is quaint, perched at the base of a steep and rocky slope that culminates in impressive cliff-tops seven hundred metres above. The buildings are from another era and display a pleasing similarity in architecture – even if one is a yellow brown and another yellow and pink. As we look down on the village from the top of the cliff, we feel sure that cobblestones will be found underfoot in the small town square.

The real-life equivalent might be on the Sorrentine Peninsula in Italy. A village situated like the marina of Conca dei Marini.

The ocean is a deep blue some days, on others it is green, but always beautiful. The road winding up the mountain was built with immense dedication and persistence to connect the village with the flat, high-lying hinterland – and with similar villages further down the coast.

Down the slope, rooted in rich, volcanic soil, grow a great many trees, interspersed with rooftops and the red and purple and fuchsia of many bougainvilleas.

Once, all our ancestors fished, and even today some boats can be seen moored close to the shore. Fishermen still go out, but economic activity has changed, we have become fishers of tourists. Buses full of visitors drive down the mountain pass looking forward to the best of pizza, a bottle of Nero d'Avola red wine, and a taste of Limoncello, all interspersed with photo opportunities.

Still, there are only 150 of us – the town will hold no more. In the mid-afternoons, the buses leave again, filled with goodwill, alcohol and even a sense of wonder, to overnight in a place where the geography allows more people. We own restaurants and work in them. We sell curios. Some of us are tour guides and interpreters, meeting the buses when they stop and walking their occupants through town, telling them about our history. A few individuals paint portraits of visitors, some demonstrate crafts from many centuries ago that have in other places been replaced by machines. Others still, more daring, perform juggling acts with fire sticks.

At the very end of town, one of the shops in centuries past was a leather tannery. It was placed at the edge of town because of the terrible smell that leather tanning produced. Mercifully, since all of us can remember, it has sold leather key-rings and sandals and wallets that had been made elsewhere.

Last summer, the owner Luigi passed away, and his daughter's husband Giovanni has now taken over the business. Giovanni was not born in our town but in the city. Last week he started up the tannery again, and ever since the sewerage-like, pervasive smell has been sold with every product in town. With the pizza, the Nero d'Avola, the Limoncello, the portraits, with the fire-sticks and with the stories of years past. The buses have been leaving not filled with wonder but with relief and a residual odour casting severe but unspoken doubt over the workings of the on-board toilet.

In the first day or two, we noticed the change but didn't discuss it much. We were busy and tried to bury our irritation. Perhaps Giovanni was doing some kind of fumigation that would soon be over. Indeed, in the beginning nobody was entirely sure where the terrible smell was coming from. But now, a week later, tempers are seriously frayed. It is from Giovanni's leather shop, it isn't going away, and it's totally unacceptable.

Three of us are at the mayor's office, petitioning for drastic action. There is much waving of arms and much gesticulation. Angry voices rise above the gentle lapping of the waves against the shore.

The mayor soon relents and calls a meeting on the town square at six pm. The three of us walk back satisfied, and as we go, we pass the message: "Town square at six... the terrible smell... town square at six... town square at six. No, Guiseppe, don't burn down the tannery yet, come to the meeting... town square at six."

Guiseppe is the town's Bruce Willis. Its Green Die Hard.

The people we tell, tell other people, and within a few minutes everyone knows.

It's summer, and by six the sun is still well above the ocean. As people walk towards the town square, they call their neighbours, and as the church bell strikes the hour, the town is there to a man (and woman, child, dog, cat, and parrot).

We gather in front of the bakery, which has a higher step in front of the shop – used as a mini-stage. The mayor calls everyone to order, and announces that there is a complaint by Alessa (who owns a restaurant). She ascends the bakery step.

"As you know," she begins, pausing dramatically and bringing her fingers to the side of her nose ,"we've been having the SMELL. Let me not say what it smells like."

"*Merda!*" shouts Edvige, and there is some sniggering and some rumbling.

"And we all know where it comes from," Alessa continues. "Giovanni, it is from your tannery, closed for about two hundred years, but proudly re-opened a week ago."

Giovanni is silent for the moment, and so she continues, " my customers no longer have a starter, or a dessert. They eat as quickly as they can, their noses pulled up, not talking, and then they get into the bus and leave as quickly as possible."

"They don't buy curios anymore!", Edda shouts. The order breaks down, everyone talks at the same time. The sun is setting quickly on goodwill and friendly relations. The mayor steps up next to Alessa and raises his hands.

"Let's give Giovanni a chance...," he says with the air of a Galileo, as if saying "let's see if the one weight falls faster than the other, the second time around."

Knowing the truth all too well – but also the demands of due process.

Giovanni walks to the front with flimsy self-confidence and the threadbare swagger of an unarmed gladiator entering the Coliseum with a lion.

"Our family has had that shop for 500 years," he starts, knowing that denial of his culpability would be futile. Aiming instead for justification. "Now we are restoring the age-old art of leather tanning. No longer will the town sell leather made elsewhere, we'll reinvigorate the fine art right here in town and sell products we have made ourselves. Our town will become known everywhere for it!"

"*Che cavolo dici.? Vai al diavolo!*" shouts Ernesto, utterly unimpressed, wishing for him to get irrevocably lost on the way to heaven.

Nicoletta, whose father had been mayor himself, feels even stronger: "*Accidenti a te!*" she shouts, imploring Giovanni to drop dead.

Guiseppe raises his hand and shouts above the din for a chance to speak. Giovanni looks at the mayor and sees clearly that his defence has run its course. He steps down, and the mayor points at Guiseppi, who starts ascending the step with some difficulty, a bucket in hand.

"Here in my right hand," starts Guiseppe, "I hold a chamber pot."

There are whistles and exclamations.

"It is a tradition that goes back not five hundred years, but thousands! All the way to the Romans. No more toilets with running water: In keeping with Giovanni's return to things of the past, I will revive the use of the chamber pot... I have done so already, since this afternoon, and the contents are here for all to see!"

He lifts his right arm with the bucket, and everyone stands back.

"As you know my father and grandfather were magistrates. What I will do to amuse tourists (and to show them how things were done in the olden days), is to demonstrate the origin of the Roman legal action the *Actio Dejecti Effusive*. Now, as you all might know, this legal action was what you instituted if you were walking down the street during daytime and someone emptied onto you, from a floor above, the chamber pot."

Without further warning Guiseppi lifts his right arm even further, adds the left and turns over the pot. Everyone at the front jumps back as quickly as they can, away from what is coming down, but they knock over the people behind them, who have nowhere to go. In a domino reaction, quite a few people are soon flat on their behinds.

"Today it was just old tea and muddy water with rotten bananas, but once I get going, IT WILL BE THE REAL THING!", shouts Guiseppi as people scramble back to their feet or try to regain their balance – half angry, half amused. "Giovanni and I will lead the town, back to the glorious past!"

"*A correre e cagare ci si immerda i garretti!*" shouts the mayor, convinced that the town has to choose between being either modern or ancient. That a mixture was sure to be disastrous. "GIOVANNI, BY MAYORAL EDICT, YOUR TANNERY IS TO CLOSE DOWN IMMEDIATELY!"

And that was that. A few more clever remarks, a lot of laughter, the sun going down and from the next morning, the smell fading away for good.

***

We often call our world of more than seven billion people a 'global village'. That is because so much these days is connected to so much else. Space as we used to understand it has shrunk.

Most of our friends can live in other countries and we can talk to them, or write them, every day. One country sneezes and all the others catch an economic cold.

And yet, we cannot snap our fingers and call a global village meeting in the global village town square. We cannot get everybody there. Not everybody present can understand everybody else. The global village is not a place of agreement and harmony. On the contrary, Marshall McLuhan, the man who coined the phrase 'global village', thought that its very diversity ensures maximal disagreement on all points.

But let's assume we use Skype, we get everyone together and we manage in some way that we all understand each other: What would have happened to Giovanni's tannery in a global village? Skype so far has no smell function. This creates rich possibilities to manipulate the message, because it will require words describing something of which most listeners have no direct and/or personal experience. Knowing that by far the great majority of fellow global villagers are not experiencing the SMELL, Giovanni would initially have just denied that it existed.

"Smell? What smell?"

This would have stalled judgment and action for a long time, for someone would have to now prove scientifically that there was a smell. If this ever happened, Giovanni could just shrug and said: "It's not the tannery – the smell's always been there".

Now, it becomes even trickier: Scientists would have to prove that the smell would be different if he stopped operating the tannery. Seeing that he was refusing to do so, this would be really difficult. Could people be interviewed who live nearby to say the Smell only started when he re-opened the tannery? That isn't scientific proof, it is anecdote.

But let's assume that with great difficulty, scientists prove that the smell hasn't always been there. Prove this, please remember, to a global village who are dialling in via Skype to the seven billion-strong global village telecon.

Giovanni now hires a retired school teacher (who has a degree in science) to do her own investigation. She concludes that the smell is coming from an old volcano becoming active again and starting to leak sulphur into the air. She points out in her report that the soils of the area are rich for the very reason that the terrain is volcanic.[87]

In the global village Skype meeting, Giovanni's lawyers assert that the scientists are divided on the cause of the smell and that the science is 'still uncertain'.

Another impasse. By now years have passed and tourists no longer visit the village. The buses no longer stop there – the tourists have no doubt there is a smell, and they don't really care where it comes from. And at the financial cost of Alessa and the other villagers who are distressed about the smell, a new investigation must be set up to prove that it isn't the volcano causing the Smell, but the tannery.

As for Guiseppi (whose best virtue is not patience), if he gets fed up with all the inaction (and the fact that he no longer has a livelihood) and burns down the tannery one night when it is deserted, he would have become a terrorist.

In a global village, there is always Plausible Deniability. People reading about a smell can never be sure that it exists, nor what causes it.

In a real village, Alessa would bear the burden of proof to show that Giovanni was spoiling the sweet smelling sea air, and would easily discharge the burden of proof in the presence of the other 149 villagers – who could after all, smell the smell for themselves (and know that it started on the day when they saw Giovanni re-opening the tannery).

In a global village, discharging a burden of proof towards seven billion fellow villagers (of whom all but a few have no first-hand experience of the situation) becomes just short of impossible.

***

But the world has to work. Decisions of collective interest have to be made even in a global village of more than seven billion people. How does this happen?

There is a way to engineer consent.[88] The information reaching people needs to be carefully controlled, packaged and communicated to ensure that the 'right message' gets through. This process was initially called 'propaganda' and more recently, 'public relations'.

The term comes from modern Latin, derived from a new administrative body of the Catholic Church, called the *Congregatio de Propaganda Fide* (Congregation for Propagating the Faith). Its activity was aimed at 'propagating' the Catholic faith in non-Catholic countries. The practice of propaganda, however, goes back as far as humanity. By 1 000 BC the Assyrians were already creating documents and monuments designed to whip up support for the king – in peace but especially in war. To show the king's prestige but more particularly to demonstrate that his actions carried divine support.

The pharaohs used public architecture to serve as propaganda of their own exalted status and legitimacy – the pyramids and the Sphinx.

Ancient Greece took propaganda to new heights, starting with the habit of calling all non-Greek speakers 'barbarians'. From there it was but a short step to any legitimate war. By 640 BC, we see the emergence of war poetry by Tyrtaeus. Religion became an important component of deciding whether to go to war, leading to increased standing for the Oracle of Delphi. If the Oracle predicted glorious victory, the soldiers would fight even harder. Alexander the Great worked tirelessly to deify himself and to keep his presence uppermost in everyone's minds throughout his vast kingdom by having his image on coins, in artwork, on pottery and on grand buildings.

The Romans understood the value of propaganda well. Cicero wrote:

> So in the beginning we must convince our citizens that the gods are the lords and rulers of all things ... for surely minds which are imbued with such ideas will not fail to form true and useful opinions.

And so it went throughout the Middle Ages, when religious justification for war was just one example of the use of propaganda.[89] By 1843 in Germany, Karl Marx was expressing the idea that something was needed to unite and console the masses. He claimed that this was religion, and famously called religion 'the opium of the people'.

In the following decades, other writers investigated 'crowd control' through information management.[90]

The century ran out and the twentieth began. The world increasingly turned into a global village, the world population grew, women got the right to vote, democracy deepened in many places. The century saw war on unprecedented scale, plus repression, revolution and persecution. In all these, propaganda played a critical part. The most relevant is how it played out in democracies, for it is in democracies that the PECCA battles of our time are not being won.

World War I was a watershed. When it came along, everyone had to decide if they were 'in' or 'out'.

The USA was firmly 'out' for most of the duration. There were German Americans and Irish Americans – not wanting their adopted country to join on the side of Britain. There were Yankees, pacifists (including Secretary of

State William Jennings Bryan), religious objectors, feminists and others. The population was edging over 100 million. Getting some kind of consensus to go to war would be really, really difficult.

President Woodrow Wilson had himself been advocating for strict neutrality. But if a week is a long time in politics, four years are an eternity. For various reasons, Wilson gradually changed his mind and came to believe that the USA needed to join the war.

In ancient times, it was easy to go to war. The king would decide and all adult men would be called up on pain of death. Agree to fight or otherwise die, now. The alternative was to die in the war, later and maybe. An easy decision. But by 1916, times had changed and Wilson had to convince the electorate that war was a winning idea. It was a gamble that could cost him the election.

Time to engineer consent. He went on a speaking tour and in 1917, established the Committee on Public Information (the Creel Commission) to promote America's entrance into the war. Journalists and communications specialists were appointed. People who understood Blackbeard's Ghost and would report positively all the time. Good news.

Good news, good news, good news. Give the people what they want. Make sure the same message reaches everyone all over the country. Censor news and publications heavily to achieve this. Use all possible mediums so that as many people as possible can be reached. Embroider the truth sometimes. Report things a little before they actually happen. Say we're ready even if we're not.

As an example of the lengths the Creel Commission went to, it recruited as many as 75 000 'four minute men.' These people were volunteers adept at social speaking and were asked to deliver short speeches at a wide variety of social events about key aspects of the war. In total they made over seven million speeches.

Along with good news about everything domestic went the demonization of the enemy. At the time, wild stories were floating around internationally about fictional German atrocities, including that they were amputating the hands of babies in Belgium to prevent further soldiers from growing up and that they were raping under-aged girls and then impaling them on bayonets, etcetera. These were shown to be completely false, but the Creel Commission seemingly joined in the merry and grandiose overstatement of German atrocities happening at the time.

Sure enough, a slow consent started to emerge and opposition to the USA joining the war waned. The country's ultimate participation from 1917 to

1918 met with far less opposition and indignation than would have been predicted in 1914, when the war began.

One of the members of the Creel Commissions was Edward Bernays, Freud's nephew, who'd emigrated to the USA as a child. He was a journalist and after the war, went into public relations. In 1923 he published Crystallizing Public Opinion. Five years later, his most famous book, Propaganda, was published.[91] It draws on the work of Freud, Le Bon, Trotter and his extensive experience in the Creel Commission and makes compelling reading.

Bernays ultimately paints a picture of a few enlightened men making decisions on behalf of the irrational masses – and then manipulating them into thinking that they are the masters of their own fate. He sees this as unavoidable in a complex civilization. He trusts in the morality of the propagandists and their masters to strive for what is right and proper and in the collective interest.

Hitler in *Mein Kampf* blamed the loss of World War I on an inferior German propaganda machine and seemingly tried his best to learn lessons from the victors – on how to win hearts and minds. So good did he become that he pulled off the Holocaust. Stalin too, used strong ideology and concomitant propaganda to kill about a thousand people a day in 1937 and 1938. There were similar purges in Mongolia and China.

In our generation, a discussion of propaganda emanated from far more libertarian authors like Noam Chomsky, who dislikes the phenomenon intensely but does not deny that it exists. Indeed, it was used to justify the 2003 invasion of Saddam Hussein's Iraq; to disarm Iraq of weapons of mass destruction, to end Saddam Hussein's support for terrorism, and to free the Iraqi people. The fact that there was no evidence that Iraq had weapons of mass destruction (none were found), and furthermore was not supporting al-Qaeda, did not get in the way of a good war, although the overwhelming weight of legal opinion was that the war violated international law.

Propaganda is part of our everyday lives, including and especially our discourse about important public policy matters, like climate change. It is sophisticated and subtle.

It relies heavily on Blackbeard's Ghost but adds lies to the information. Untruths. Error messages. It says, "Know that your audience will not evaluate like a computer, knowing only zeros and ones. Play to their prejudices. Let these prejudices do the work for you."[92]

***

As the yacht drifts inexorably downstream under cloudy skies, separated from the reassurance of the stars and the ever-same constellations, the world drifts downstream on matters like climate change[93] and biodiversity loss.

As Juan-Franco and Julia toss and turn many scientists and activists sleep badly at night about climate change. Just as Julia has put up the science on the mast for all to see, the science about climate change has been made available for all to see and read. On a UN-sponsored website. Like Julia's document, practically nobody reads it. People continue to lie in the sun, read their i-pads, drink Mochitos.

And as the yacht has a point of no return, the solutions that science suggests to climate change have a sell-by date, after which they are worthless. After which our civilization and our world are pulled irrevocably into the eye of the cyclone.

In the field of PECCA's, science and propaganda are involved in mortal combat for the hearts and minds of the general public.

Science aims to win minds. Propaganda only aims to neutralise minds and then to win hearts. The science of cognition shows us that in the short run, propaganda has the more scientific approach – we do better by winning hearts than we do by trying to win minds.[94]

This unscientific approach by Team Science is surprising. A scientist would never take the Lunar Roving Vehicle used in the Apollo 15 to 17 missions to the moon and send it to traverse Mars without looking for every possible way in which the two environments might differ. Because the slightest difference might mean the vehicle designed for the moon would not work on Mars.

And yet, astonishingly, science uses exactly the same approach to communicate its findings, irrespective of whether the audience is one person or the global population; heedless of whether the audience is wanting to hear the message or defending against the knowledge; regardless of whether the solution can be implemented without majority approval or not. Science treats everything the same. Selling science is not part of what scientists do.

The equally scientific research and experiments that have been done about cognition and the way people perceive are completely disregarded. In striving for the objectivity and empiricism demanded by the scientific method, science has become ideological. Because it believes that all scientists should be cold, rational and objective, it point-blank refuses to make any allowance for a general public that is neither rational nor objective. Science serves a vegetarian meal to a restaurant full of carnivores.

Blackbeard's Ghost takes a graph plotting truth and lies, beauty and ugliness and distorts it by dragging 'lies' higher than truth. Our human prejudice for good news creates a built-in advantage for lies to be compelling, as long as it's part of concluding 'Everything will be okay.'

Just there, the science surrounding PECCA's breaks down, because in negotiating the last and most critical link of the chain – communication - it refuses to be scientific. Instead it prefers to be dogmatic.

Just as Heisenberg put the observer back into the experiment, science needs to put the audience back into the communication of the knowledge.

But even that may not be enough. There is a sell-by date to scientific solutions. Science needs to win, and win soon. It needs to solve the problem of the armed guards at the door of perception.

Science needs to find the back door of perception, and get in where there are no armed guards.

23

Juan-Franco and Sam

It's somewhere near midnight when Juan-Franco comes out of the shower and reaches his bunk. Across from him, on starboard, Erin lies in the other bunk, seemingly asleep. Lorenzo and Angela are in the other double suite towards the stern, across from the one occupied by Sam and Angela.

The day has been long and full of all sorts of unexpected developments, and Juan-Franco feels really tired. The work on the improvised sea anchor has strained his physique. Too often he tends to forget that some things he just can't do anymore. Simple things. And everything else takes longer than it used to do before the accident.

He lies down, ready for sleep and yet with much on his mind. It will be six or seven hours before he'll have a chance to talk to the others. How to approach them? Anger or fear-mongering won't work unless they can see for themselves what's happing with the weather. They haven't read Angela's notes on cyclones, and won't have the benefit of the explanation she's given him earlier the evening on the deck. All they know is what they heard from her earlier, when she was anxious and when Sam was doing his utmost to steer everyone's thinking in another direction.

Sam is key. For some reason he just won't hear of a cyclone. Juan-Franco realises that he should have asked Julia if she knew why he's reacting like this. Sam is always too sure of himself, always somewhat disdainful towards just about everyone else – but this seems worse. If Sam can be brought onside, it'll be easy. They can even use the engine to help them steer in the direction Angela wants them to go. Away from their destination, sure, but what does that matter at a time like this?

He rolls over in his bunk, but sleep still evades him. What could it be with Sam? He recalls the exchanges following Julia's interventions, and tries to distil the common denominators. Has there been earlier tension between

Sam and Julia, and he wants to humiliate her? If this is so, his opposition was aimed at the messenger and not the message. But no, there isn't moodiness or spite in his demeanour, there is cool calculation. It has to be something else.

Again Juan-Franco regrets not asking Julia about it. But then, there isn't a script or a textbook for days like the one that's just passed. Well-intentioned actions are taken, and snap decisions are made. Some turn out to have been inspired, others not so. In any event, Julia might not know the answer herself. Probably doesn't know, or she would have volunteered the information. The level of her surprise and disappointment at Sam's conduct indicated that she can't explain it – other than by concluding that he is a much, much bigger jerk than she's ever realised. Which of course he is and always has been.

But it doesn't necessarily explain why he was so utterly dismissive of Julia and her warning. He was worse than just difficult, he was unplayable.

Well, start with the others then. Just three people. Lorenzo and Angela are very close and unlikely to go in different directions. They would decide together.

Erin seems to have the capacity to make a stand once she's sure about her own thoughts. She is very much in holiday mode, but she'll soon snap out of it as the weather deteriorates.

The question is not only one of what to say, but how to approach them. Especially as Sam will be everywhere, trying to block conversations. So there probably won't be much time to explain as Julia had to him on the deck, after they swam. Juan-Franco knows he'll have to either start really early while the others are still inside and can be spoken to one-on-one, or if that proves impossible, that he'll need something that can be said, understood and agreed upon within thirty seconds.

He took off his shirt just before sunset, and forced them to see what it looks like when a body collides with a reef. Deliberately shook them and then followed that immediately with evocative guitar music and singing as the sun went down. They would remember this in the morning, would treat what he was saying differently to the way they treated Julia's words earlier on.

The back door is ajar. What is needed is the right message. But he doesn't have it yet. He'll have to sleep on it.

He feels the movement in the hull. The lifting stern as the swell catches up with the yacht. Then the mid-ship, where he is, lifting. Then the bow. Almost no roll, it has to be coming straight from behind. He counts the spaces between the swells, habitually moving his left foot like he always does to keep time, when he plays guitar.

Seven seconds. Seven seconds. Six seconds, seven seconds. And the swell is high. Even lying in his bunk he' sure it's higher than before. He listens for the wind in the rigging, and hears it clearly. Is it raining yet? Doesn't sound like it. But it's all exactly the way Julia said it would be. They're being dragged towards the eye, which for the moment is likely to be moving away from them. It might move faster than them and leave them behind, but that doesn't seem to be the case at all, or the frequency of the swell would be decreasing, not increasing. And the cyclone might change direction or even reverse its course, in which case their end would be racing towards them.

That is tomorrow's battle. He turns over one last time to find a comfortable position. The humidity is stifling, and he reaches for his still-damp towel and lays it over the pillow. Deliberately slowing his breathing, he finally falls asleep.

Just like Julia on the deck in the afternoon, he's failed to plan and schedule his next move clearly. And Sam is lurking.

***

Sam slept very lightly. There was a lot to be done, and he had to be ready when the time came, or it would all be gone.

He'd wake up often, lie quietly with closed eyes, think about the hours ahead. First prize was still to out-manoeuvre everyone. That way, he'd get what he wants without alienating anybody, without inviting legal proceedings that could be avoided, without feeding the rumour mill.

Occasionally he'd glance down at his expensive watch, and push the button illuminating the dial. Just to be sure. Two o'clock. The next time he looked, ten past three.

The question was how to approach them. He had to catch them inside, one on one, without waking Juan-Franco or Julia. By the time they wiped out their eyes it had to be a *fait accompli.*

Sam, aware that he's dozed off, looks at his watch again. Quarter past four. The sun would rise at five thirty – he checked before going to bed. So first light at say five, or just before. He relaxes for another fifteen minutes, then glides out of bed and dresses quietly. He puts on a loose-fitting cotton shirt, retrieves the revolver from between his underclothes and socks where he's hidden it, straps it on below his left arm and buttons up the shirt quite high to hide it from view. Insurance. An oilskin jacket would further camouflage it.

He then goes outside. Yes, the wind is much stronger, must be twenty-five knots or more, and the swell has grown. Not a cyclone. But quite a little storm. He walks to the mast and finds Julia's dissertation in its plastic sleeve. Taking out a multi-purpose pocket tool, he opens the blade and cuts the cable tie.

With the plastic sleeve in hand, he steals back inside and without knocking, enters the suite where Lorenzo and Angela are sleeping. He couldn't knock – Juan-Franco and Erin are sleeping very close by and would surely wake up. The wind is whistling through the rigging now, drowning lesser sounds, and he feels confident he can lure Lorenzo outside without waking everyone else. He goes to Lorenzo and touches him on the shoulder, firmly but calmly. Not to frighten him, but to assert some authority.

Lorenzo wakes quickly and moves to sit up. Sam puts his hand to Lorenzo's ear and whispers: "Wind has got up, we have to get going. Please dress and come outside, and I'll explain."

He doesn't wait for an answer, doesn't give Lorenzo a chance to argue. Three minutes later he emerges on the deck, sees Sam waiting for him. It's getting light, and Lorenzo can distinguish large swell with whitecaps. Conditions are getting filthy. Sam puts his hand to Lorenzo's ear again, to combat the wind and not shout so loud everyone wakes. The plastic sleeve is in the other hand, and he waves it as he speaks.

"I've studied this, and she's right, we're in a low pressure system that gets much worse in other places. It says we have to run with it, not fight it, and then start beating after about four hours. We have to use to the wind and swell to get away from it and move to starboard a little later when the swell dies down."

Lorenzo, slightly sleepy and slow, nods. Not fighting sounds right. Getting away from it sounds right too. And Julia knows what she's talking about, so if this is what she says, then good.

Sam continues: "Julia and Juan-Franco worked all night on a sea anchor, should we need it. They are absolutely beat, so I don't want to wake them now. If you could quietly fetch Angela, we can get going very soon. Go, please, we have to hurry!"

The urgency in Sam's voice evaporates all residual drowsiness and Lorenzo does exactly as he's been asked to. He quietly wakes Angela, explaining first that according to Julia, they need to run with the wind. He continues to tell her how tired Julia and Juan-Franco are and that she needs to come outside. "Looks like rain," Lorenzo says in her ear, "bring our oilskins."

Less than 10 minutes later all three are on deck, and there is sufficient crew to implement Sam's plan. Raise the spinnaker, trim the sails, make a hard run dead downwind. North-northeast, with the island northeast. They'll run with it. And Sam firmly believes that the swell will gradually ease, as will the wind, and then they'll beat their way east – in and beyond the reefs from there. Use the engine once they're close. Getting close would take no time at all – not with the wind blowing the way it is.

For Sam, this rosy outlook is not a lie, it's his version of reality. He believes it because he has to. Any alternative leads to unthinkable results.

What was a lie was the way he represented Julia's facts to Lorenzo and through him, to Angela. But that was a means to an end. To Sam, this lie was essential to ensure that the others make a responsible decision, because they don't generally know what's good for them and they don't really understand what's happening.

And so, the big, fat, ugly, propagandistic lie; Julia says this is what we must do. Julia says if we do this, we'll be fine. And not just Julia, Julia the scientist. Science as per the document in the plastic sleeve that Sam just chucked overboard ten minutes ago.

According to Sam, the crew on the yacht and the crew on planet earth are doing what science tells us to do when confronted by a cyclone or impending menace; raise the spinnaker, trim the sails, full speed ahead.

Everything will be okay. Everything will be just fine.

PART 4

Beautiful Truths

24

NERO'S FAMOUS LAST WORDS: "WHAT AN ARTIST DIES IN ME!"

Rome and Antium, Summer, AD 68

It was 17 June of the year 68 and Nero was retracing his steps.

He'd been born in Antium, a few miles outside Rome, and had moved first to the capital, then to the pinnacle of his civilization – if it could still be called that. And now everything was going back the other way. From Rome, back to Antium. Dust to dust. Nothing to nothing.

The anxiety was like a python around his neck, strangling him.

They're gone. All gone.

All his friends, even his enemies, all the gladiators. Every one of them, gone. Poppaea, Seneca, Petronius, Britannicus, his mother Aggripina, all gone. And all the others, the ones still alive. The friends to extradite him, to get out of the way for the inevitable, not to be crushed with him. The enemies to aggregate and seal his fate and to be able to say they were right in the middle of his dethronement.

The carriage bumped along over the cobblestone road, and Nero felt like vomiting. For a moment he fantasised that it was some kind of motion sickness, but then he remembered the scene at his palace just an hour before. How he'd run out to throw himself into the river, and cried; "Have I not friend nor foe?" How the inane silence had taunted him, whispering:

"Yes, yes, yes, YES! You have nobody. You are all alone. No friend, no foe, just you!"

He remembered again how his legs had refused to carry him into the river and had stopped on the edge, his words echoing off some wall and coming back to an utter coward who had to skulk back into his empty palace, wondering what to do next.

It wasn't motion sickness, it was fear. Utter dread.

How did it happen, how could it happen, where did it all go so wrong? Was there something he could do, some little bit of hope somewhere? Perhaps voluntary exile and the governorship of Egypt – perhaps then they would spare him?

The carriage shook, as if even the horses were picking the roughest route they could to defy him. He got up to order that they be thrashed, sensed the futility, fell back in his seat. It was dark, with no moon. He remembered that his party were all travelling in disguise, that he shouldn't be making a fuss.

Less than four years before, in the late summer of 64, everything had been perfect, the fire had cleared more space than ever he imagined it would, Petronius was churning out one brilliant idea after another for the new palace, and the plebs were loving the musical renditions he was starting to give with his cithara.

Then came his triumphant musical tour around the empire, as far afield as Athens herself. Everywhere he was received with overwhelming enthusiasm. The cultured Emperor, the Athenian amongst Trojans. Then followed the inaugural musical competition covering the empire, the Neronium. Which he won, the gods be praised.

And his statue was going up on the Via Appia. The colossus of Nero. The biggest ever in any civilization, a hundred and twenty feet high. Not equal to the achievements of the man it was immortalising, but fabulous nevertheless. And Poppaea was finally pregnant, so an heir was on the way.

But just the next year, the small-minded bastards got their chance. Gaius Piso and his conspiracy with Lucan and Plautius Lateranus and ever so many collaborators. Trying to kill him – Nero – ostensibly to save the empire from his populist tendencies.

The threat on his life had been annoying, but not unexpected. In the decades since Julius Caesar, much had changed. Caesar who'd defied the rules of the Roman Republic and started a civil war by marching his army into Rome, emerging as the undisputed 'dictator in perpetuity'. The careful conventions of the Republic had been torn up and politics became increasingly marked by intrigue, conspiracy and murder. So a plot on the Emperor's life was normal, and could be dealt with in the normal way; By executing the guilty as swiftly, violently and painfully as possible.

What caused Nero to fly into an apoplectic rage was the lack of respect for what he was doing for the Roman Empire and for history. Lucan, a poet, who'd seen Nero singing poetry to the masses, accompanied by his cithara. Of all people Lucan should have seen that culture was returning, that a golden age was dawning. But he was blinded by jealousy, or so thought Nero. He, Lucan, wanted to be the one singing, and being applauded. He wanted to be the one winning the Neronium. So Lucan was executed, with Piso and Lateranus.

If only that had been the end of it. But no; the rotten apples had to spoil the good ones. Seneca, his mentor when he became emperor, admitted that he'd talked to the conspirators before. And had to be forced to commit suicide.

There was no other option. And that was Seneca gone.

Much worse: Petronius was also implicated in treason. Petronius his soul mate in refinement and sensuality, the court's arbiter of taste. And had to be forced to commit suicide.

There was no other option. And that was Petronius gone.

Nero was highly irritable after that. People stayed out of his way, the clever ones who knew he couldn't control his rage. But not Poppaea. She had to stand in front of him with her huge pregnant belly and wave her finger at him for spending too much time at the races. And so he kicked her in her stomach until she shut up.

There was no other choice. And that was Poppaea gone and the heir gone.

And then Piso and Lucan. All the people he could have turned to on this darkest of nights. If they hadn't been corrupted, they'd still be alive and he, Nero, wouldn't be alone.

But it didn't start going wrong with Piso. The economy had gone into decline with the rebuilding of Rome. Not enough taxes could be raised to bring Rome to the required levels. To fill in the ugly marshes, to build the new palace, to dig the canals to bring water to ugly landscapes. And to fund the expeditions into Africa, to find the source of the Nile.

History demanded that all these things be done, as Rome's contribution to posterity. To show why the Colossus of Nero needed to be so high – and higher still. And the utterly uncultured provinces responded by whispering that he was ruining them and started all the resistance in Germania and Hispania, led by Galba and Verginius. Low scale civil war ensued, with the Emperor's position getting weaker all the time.

Until, suddenly, it was all out of control, to the point where an officer who a few years before would have feared him more than the gods could suddenly refuse to help him escape by sea, and say sarcastically, "Is it so dreadful then to die?"

An officer, his officer. A nobody innumerably far below him in the natural order of things. And then, tonight, waking after fitful sleep to find his palace deserted. Everyone gone. Not even a gladiator to kill him as he called out for a merciful end. He was the last to know that it was all over, and suddenly he was down to three loyal freedmen and his secretary, Epaphroditos. And travelling in disguise in his own kingdom.

They were coming for him.

***

The anxiety was strangling him. His chest just wouldn't expand to allow breathing, and he was sweating even more than the summer's night could account for. He thought of the many nights he could have had anything his heart desired – and did. Now, all he desired was a little hope for a miracle. But he couldn't see where it would come from.

They would arrive, horsemen, to take him away. There would be a lot of pain, and then the end. After everything, it would end worse for him than it did for Caesar. Again he asked himself how it could have happened. As if it mattered.

They arrived at Phaon's villa, and Nero disembarked. Would he ever be in a carriage again? Would he ever retrace these steps he was taking now, left foot, right foot, leaden left foot, heavy right foot?

"Dig me a grave," he half ordered, half begged. His voice was breaking into the same falsetto that had rung out four summers before, when he sang *The Sacking of Troy* while Rome came alight. "Look for marble to line it."

The freedmen had no marble, had never seen marble, had no idea where marble comes from – but they dug away without arguing.

Then came the sound he'd been expecting, but it was only one horseman, coming fast. It wasn't the mob, nor the centurions. Not yet. It was a courier.

Nero stepped into the light, opened then read the sealed note, then pictured the impending scene as he'd pictured the Golden House; vividly and in great detail. The Senate had declared him an enemy of the State and had decreed that he should be arrested and beaten to death in the Forum.

He played it out in his always superior imagination; the flickering, yellow-orange light of torches and lanterns, the long, eerie shadows, the raising of arms bearing heavy and blunt objects, the cowering victim, the coming together in a circle, the frantic blows. The heap of broken bones, pulverized flesh and body fluids as they step away and back into their dignity.

There was only one hope, and that was to avoid that ending. To go somewhat more mercifully, a little less humiliatingly.

But he'd frozen on the edge of the river, and he froze again and again. Sword in hand, he couldn't bring himself to fall on it.

"What an artist dies in me!" he whimpered as he started to cry and put his sword away again. "What an artist dies in me."

It was all over, all the glory, and was just blood and guts. For the moment no guts and no blood. But that's what it was.

Then the many horses could finally be heard, approaching more leisurely. They had no rush, for the emperor had nowhere to go. They were doing their jobs. Fetching the condemned like they'd done before, when he'd been issuing the orders.

He took his sword out again, the grave diggers froze as they watched him. Again unable to do it, he cast it away with a cry – his chariot to the next world. In a panic, he implored his secretary to help him.

"Kill me, Epaphroditos! I beg you, they're here!"

He saw then definitive intent in his long trusted, right hand man. The horses were close, and Epaphroditos had little time to dissociate himself from the emperor and survive. As he took out a knife and bore down on Nero, the cry came louder than before, "*Qualis artifex pereo*! What an artist dies in me!"

Seconds later his chest had been ruptured and blood was spurting all over. He was barely alive as the cavalry arrived and he exclaimed once more, much softer, with halting breath: "*Qualis artifex pereo.*"

What an artist dies in me.

And then the artist was gone, spoken a true word as his last, after a life of disdain for the truth.

For if art is the elevation of the captivating over the commonplace, then Nero was an artist through and through. He incorporated it in everything he did, from playing music to the masses, to the way he destroyed a city he disliked, to the way he rebuilt it, to the way he murdered people.

Art was life. Life was art. Neither imitated the other – the distinction had been erased. There was no truth, no falsehood, no right and no wrong. Just art.

All art.

25

Rude awakening

As Juan-Franco slept, he was subconsciously checking the swell and wind for change – like a farmer with nearly ripe fruit checks for an oncoming hail storm. Everything depends on it, and while he probably can't do much till morning, he has to be sure their sense of timing is more or less correct. If Julia is right, first light would show much higher swell, much stronger wind and perhaps fits of heavy rain.

By morning, it would be more difficult to broach and take a few swells broadside while bringing the yacht about to a south-easterly close haul – but it would still be do-able. The advantage of waiting till morning was that there would be little argument by then that Julia was completely right in her prediction of the unfolding weather. Secondly, it would just be easier to pull off that first, dangerous move in daylight. They can, if there's any doubt, avoid having wind and swell both threatening to capsize the boat to leeward by using the engine and hoisting sails only once they're heading very close to the wind. Or even motor dead into the wind until the fuel runs out, just to get away and into calmer waters.

But that was if she was right about everything – most importantly right about the dual assumptions that they're in the trailing quadrant of the cyclone and that it isn't moving towards them. The conclusion that they're being carried closer at moderate speed by the wind and swell, because the cyclone is also moving away from them. The deduction that conditions would deteriorate more gradually than they would have had they been stationary on land, with the cyclone approaching at perhaps five knots an hour.

Listening and feeling, he slept for brief periods, lay awake again, drifted to and fro in a nebulous semi-consciousness. If at any point a squall or rogue wave arrived, he would have been wide awake and on his feet immediately, ready to call everyone into the saloon and to urge them that they couldn't wait till morning.

But the change was gradual. Swell grew incrementally, the frequency increased slowly, wind rose marginally. All bad news, but expected and in line with Julia's predictions. He thought about Esthia, or perhaps dreamt about her. She had to be thinking about him, worrying and wondering what she could do. He wondered if she's written an email and reminded himself to check when they had 3G reception again.

It's still dark below deck when Juan-Franco realises that something fundamental has changed in the movement of the vessel. All through the night the swell ran through from stern to bow every seven seconds, rocking the yacht and her inhabitants like trusting babies in a cradle. Sleep, sleep my darlings, Mother Nature is here and loving you and won't let anything bad happen to you. Sleep, sleep and trust my love. Everything will be alright. Everything will be just fine. Hush, sleep.

But now it's different, the yacht is stirring and coming alive. The regularity of movement is gone, there is more roll added to the pitch, the frequency of the swell seems disrupted. He sits up, tries to understand what could be happening. There is a jerk forward, and a simultaneous heeling over to port.

They're sailing. Someone has taken the initiative already – but not with full crew. Across to starboard, he can see Erin lying in her bunk. So four at most on deck. If they're heeling over to port, starboard was windward and they had to be heading further east than before. All good, but if they're coming round to southeast, the heeling would increase dramatically as they broach and reach before beating. Which means the most critical move was imminent and he has to be on deck. Everyone have to be on deck. Why didn't they call them? And why didn't they use the engine to come about before hoisting the sails, it would keep an even keel and lessen the effect of taking the swell broadside in the turn? Why toy with capsizing?

He tries to get up against gravity, but as he does so, the heeling decreases and the even keel is restored. The sails must have been lowered again, or the direction changed to run downwind. The first possibility would be good news if they're moving to the engine to come about. The second would be utterly insane – chasing the cyclone until catching the eye.

For a moment he tries to find his clothes and shoes in the dark, then he dispenses with civility and switches on the light. "Erin!", he calls, "Wake up! We have to go up!"

She stirs, sleepy and confused. He dresses, still monitoring the movement of the hull. Juan-Franco finds his shoes: cross trainers, well worn, and sits down to put them on.

He's been listening to gauge the strength of the wind, and it's obviously considerably stronger than the night before, drowning any minor sound in the saloon and cabins. Now, though, another sound starts to emerge, one that he's been told about by experienced sailors but has never heard himself; the hull starts to hum. It's oscillating and snarling, and this can only happen at pace on a dead downwind run. They're under far too much canvas and moving at great speed, and the hull's planing rather than ploughing through the water. The back of the boat is probably lifting, and steering thus almost ineffectual. At any moment they could broach or worse, the bow could bury itself in the bottom of a swell and the boat could somersault like a bicycle when the front brake is pulled hard at high speed.

Juan-Franco feels himself losing his temper as he fastens his laces and gets up. It's Sam. This can only be Sam, chasing a cyclone at 20 knots per hour. Four times the speed at which the cyclone is moving. And as the swell grows and the wind strengths, they'll start catching up ever faster and all the time turning away, broadside to the waves, would become increasingly impossible without capsizing.

At that moment, Julia's cabin door swings open and a shadow emerges hastily. It's been getting light. It's Julia, and she doesn't say anything to him or Erin. She's swearing incessantly as she beats him to the steps leading up from the saloon to the cockpit. She storms up and out with Juan-Franco in close pursuit.

As he puts his head out, she's shouting at Sam already, at the top of her voice, but Juan-Franco's shocked to hear how small she sounds. The wind is roaring like 80 000 people clapping at a stadium rock concert. The swell is wild and enormous, and spray's flying off the tops, coming across the deck like a horizontal thunder storm. Sam is at the helm and Julia's in his face.

"ARE YOU OUT OF YOUR LITTLE MIND!" she shouts, pushing her face close to his. She pauses for a moment, not sure how to follow up from such an opening. Juan-Franco turns, sees the spinnaker flying, looks at the silhouette of the mast and feels sure that he sees it bending under the immense strain. They're moving straight downwind, the boom way out to port, and they're going so fast that they're overtaking the swell. As they cross the fourteen- foot breakers with their white caps, the yacht is surfing down the other side, stern aloft and bow seemingly sure to bury in the base of the next wave and pitch-pole, but then miraculously climbing again as Sam steers slightly to the side each time to lessen the angle.

The day continues to break, but it's overcast and still hard to see detail. Lorenzo and Angela, hair drenched, sit at the side of the cockpit. They must have been responsible for trimming the sails and flying the spinnaker – now there was little for them to do. They watch Julia still shouting at Sam: "THIS IS A CYCLONE ... WHAT PART OF THAT DON'T YOU GET? AND YOU'RE CHASING STRAIGHT INTO THE EYE! HOW ARE YOU GOING TO TURN OUT OF THIS?"

Sam too is drenched, but his jaw is firmly set in a self-satisfied grin, and his eyes large and transfixed in an incongruous ecstasy. Man as master of the world, taking the worst nature can come up with, and staring it down. Hit me with your best shot. Which is more or less how he reacts to Julia.

"I'M DOING EXACTLY WHAT YOU SAID IN THAT THESIS YOU TIED TO THE MAST...", he shouts back without taking his eyes of the sea for more than a split second at a time. "HEADING NORTH AND USING THE WIND TO GET AWAY FROM THE LOW PRESSURE SYSTEM – IT WILL EASE BY ITSELF AS WE MOVE AWAY AND WHEN IT DROPS COMPLETELY WE'LL USE THE ENGINE TO GET BEYOND THE REEFS ..."

Sam sees her drawing a deep breath to come again. Before she can do so he takes his gaze away from the chaotic water mass, looks at her with complete focus and startling menace, and shouts as if to the most despicable creature alive: "NOW STOP GETTING HORMONAL WITH ME AND GET OUT OF THE WAY!!"

An argument, an altercation, even a loss of concentration of just a few seconds would see them somersault. Juan-Franco grabs Julia's wrist with his good arm, and squeezes hard. She's right, of course, she's been right all along, but insanity is at the helm and a direct confrontation now will see them drown before the sun is up. He knows he's probably hurting her, but she has to stop and retreat immediately, and find another way.

He feels the tension go out of her arm as she too, realizes the futility, then tugs gently at her arm to prompt her to come back down with him. Her posture changes, the aggression towards the husband she's come to detest so much dissipates, defeat creeps in.

Sam, knowing he's won another, perhaps pivotal round, has his eyes on the next breaker – bigger than any of the ones that went before. They ascend, up and up and up, like that breathless climb to the top of a roller-coaster ride, then balance at the very top for a mini-eternity. Visibility is up to maybe 300 feet now, and the sea is black, blue, voracious, with white caps and anger everywhere. Un-buffered, the wind roars even loader as the spray abates for a moment. Juan-Franco, blown over, takes a small step to not fall over. Then they're roaring down the far side of the wave, into another valley of the shadow of death, the bow almost immediately reaching the foot of the following wave and needing to climb, climb, not bury, not stick.

So it would go as they chase into the eye of the cyclone at a net speed of perhaps 15 knots an hour. The waves would get higher and closer together and the wind stronger, the angle of ascent and descent more acute. In Sam's mind, it would all dwindle to the point of being becalmed, and they would start the engine and motor in over the last stretch of mirror-like, friendly seas. But in the real world the only question was how close they could get to the eye before perishing.

Time for urgent action. Juan-Franco tugs again at Julia's arm. Just then he catches Angela's eye – scared and bewildered. He gives her the slightest of smiles – just enough so she knows he knows what she's feeling. Turning, he starts to descend the steps again, holding on to Julia's arm as long as he can reach. She takes the cue and follows.

At the bottom, Erin stands finally dressed and ready, with big eyes that are wanting to know what's going on and why all this shouting.

Then Julia is there too, and looking at Juan-Franco she sees determination, urgency and intent and feels an overwhelming relief that someone still has a plan.

26

THE BACK DOOR OF PERCEPTION

Just a day earlier they were on the deck in perfect conditions and Julia, randomly skipping past channel 16, heard a storm warning on the VHS radio.

Not once, but three times she emerged on the deck to convey the gravity of their situation. The first time, she was unarmed, trusting in the goodwill of her husband and fellow sailors and friends. By the third time she was, as a scientist, armed to the teeth. She came with facts and figures that would do the talking loud and clear, that would stridently persuade. That would obliterate any doubt and would galvanize action. So she believed. So she assumed. Erroneously.

Julia was trying her utmost to convey something extremely important to other human beings; there is a cyclone coming! We're in this together. We have to do something, and it's different to what we're doing now. The Mochitos and the i-pads and the sunscreen must be put away for another day.

She'd never call her account a story – but her audience hears countless stories dished up as fact. So many that their world is approaching one where nothing is real. The weary public, bombarded with lies dressed up like truth, has assumed a default position of "everything is false". Every alleged fact is assumed to be fiction. And so, while she may not recognise it and would never approve of the notion, Julia was standing on the deck engaged in storytelling.

The storytelling of science. On that day on that yacht and in those circumstances, perhaps the most important storytelling there can ever be, for she is the sentry that has seen the approaching danger and has to warn the villagers or the co-sailors that there is a wolf or a cyclone coming, as the case may be.

She's a latter-day scientist, standing on the shoulders of the giants that came before. Communicating as knowledgeable people have done for at least 23 centuries. Julia (even though she is excited by injustice and stupidity) communicates only cold, hard facts. 'Warm facts' is not a term that exists in her lexicon, much less the concept of 'beautiful truth'.

In the time of Aristotle, this wouldn't have mattered, for the term 'scientist' didn't exist, and wouldn't for another 2 000 years. Aristotle wrote about poetry with the same enthusiasm he used to discuss physics and politics. Likewise, theatre and music received the same attention and respect as logic and rhetoric. There was one reality that he lived in, and he wrote about all of it.

The tradition was upheld and enhanced some 500 years later by Ptolemy, the great Egyptian, who wrote about mathematics, astronomy, geography and music, and composed poetry.

By the time Leonardo da Vinci was in his prime 1 400 years later in the Renaissance of the late 15th century, it had become the vogue for humankind to embrace life in all its facets and in all possible depth, not just collectively but individually. While Aristotle wrote about all of life and Ptolemy dabbled in poetry, Da Vinci was all things; painter, inventor, musician, mathematician, sculptor, anatomist and much more. There wasn't even a suggestion of a schism between his scientific self and artistic self – it was all one. If it had been Leonardo rather than Julia talking about the cyclone, the rendition may have been different. May have approximated 'warm facts' or even approached beautiful truth.

But times changed after Da Vinci. In Newton 200 years later, we already find a dispassionate genius who hardly seems to think of the arts. The universe to him is a giant clock made by the Supreme Being, everything ticks with complete predictability. If we know where each atom is and where it is heading, we know the future of the universe as surely as we can predict the movement of billiard balls on a table. This world of Newton's is already hard in its programming, and cold in its predictability.

Then the Enlightenment came, the Age of Reason. It was all about order, structure, balance, reason over emotion. Newton's universal clock remained the master metaphor. Truth became cold, and facts became hard. And increasingly, it was expected of clever academics to be perfectly rational in all aspects relating to their work. The silent understanding was that science was the primary place where rational people came to interact, at a higher and purer level than everyday life.

For Aristotle, all of life was one, but by the late 18th century, there was messy life on the one hand and on the other the separated reality of pure rationality, an enclave of perfection. An embryonic space within which to cultivate the future uber-man.

Another 50 years later, the divide would be complete, for a name would be invented to cement it. It was the English historian William Whewell who remarked that even within science, knowledge was becoming fragmented, and the term "philosopher" no longer described all the different and increasingly distinct disciplines that were arising in the pursuit of knowledge. Thus, the advent of "scientist", ironically with reference to the word "artist".

An Artist is a Musician, Painter, or Poet, a Scientist is a Mathematician, Physicist, or Naturalist.[95]

And there it was, opposed. Juxtaposed. Artist and scientist. Leonardo split in two like the atom. The inseparable, divided.

And of this legacy was born Julia, the scientist.

Julia is of this world in its totality. She loves her daughter and fresh doughnuts with custard. Sometimes, she's afraid of the dark. When she sees a swallow or a woman with flowers in her hair, she often starts to cry. And when she talks about world hunger she gets angry and her cheeks get flushed. But when she communicates about truth, she dons a different persona – or tries her utmost to do so. She only talks about the cold, hard facts.

Which is okay. Some people kiss on the cheek when the say hello, others on the mouth, others shake hands. Conventions differ. If the majority of scientists wish to deal only with facts that are cold and hard, they're free to do so. For cold and hard truths are still vastly preferable in the advancement of our world to any form of lie – beautiful or ugly.

Two problems arise. The first is when the scientist is communicating with other scientists who for some psychological reason cannot accept the truth at that moment in time. Perhaps because it would make the listener complicit in the deaths of hundreds of people. This is the problem Ignaz Semmelweiz had.

The second problem in the storytelling of science as it has evolved is that someone like Julia doesn't always communicate with people who accept, understand or practise these conventions of supposed, cold rationality. Like the other five people on the yacht. None of them is a scientist, and while all are of capable of rationality, it is not a natural default position for any of them. They can be brought to rationality if Blackbeard's Ghost doesn't prevent it, but it isn't second nature to them in the way that a cigarette is second nature to a smoker in the morning.

Their rationality as listeners is acquired. It's like rain; in every place on earth, it rains sometimes. But nowhere does it rain every day, all day.

There was Nero and the beautiful lie of the violin being played against the backdrop of a city ablaze. The beautiful lie is all about captivation, appealing to our need to be transfixed, inspired, elevated, mesmerized.

The ugly lie has no such pretence. It's perfectly content to appeal to the wide, base cracks in our rational armour; suspicion, paranoia, xenophobia, myopia, superstition. It was most often the ugly lie that fuelled the holocaust and the purges of Stalin, the burning of witches, and the persecution of scientists. Over the years, the decades, the centuries and the millennia. It was the ugly lie that justified many of the senseless wars, the mindless killings, the pogroms, the Inquisitions and the genocides. And it was against the reign of the ugly lie that science rallied with Rationality and the Age of Reason. With the advent of cold, hard facts.

But science and truth have not yet developed a weapon against beautiful lies. Anything occasioned by Blackbeard's Ghost falls into this category. We have a vision of the world as it should be, and in it we are happy. For Sam, this means money. For the doctors and academics opposing Ignaz Semmelweiz, this meant respect. For people hearing about climate change or biodiversity loss, it means a happy and sustainable planet not just for them, but for their children and their children's children.

If we as scientists confront ourselves as humankind with something that ends with, "It's all over folks, unless you change everything about the way you live the world is going to the dogs, and we're all going to die", we cannot hear that truth. It is ugly, in the way that the ending to *Atonement* is ugly.

> How could that constitute an ending? What sense of hope or satisfaction could a reader draw from such an account? Who would want to believe that they never met again, never fulfilled their love? Who would want to believe that, except in the service of the bleakest realism?

Such an ending lacks appeal just like the book version of *Captain Corelli's Mandolin*, where the lovers lose all but a sliver of their lives together and only get reunited in old age.

And it is ugly in the way that Ignaz Semmelweiz's assertions were ugly, saying: "You clever doctors are killing the patients with your dirty, smelly hands. The mothers and the babies. Hundreds and thousands of them."

Every time, the seesaw comes down the wrong way. The weight of evidence seems to remain suspended in the air, the flimsy tatters of doubt on the other side hold sway and go to earth. Because of Blackbeard's Ghost, invisible on that side.

In looking for a way to overcome this unbalanced outcome, it's worth noting how science approaches its battles. There is no strategy, no tactics, no subtlety. By choosing to express itself in dry and precise language devoid of adjectives, metaphors or beauty, science has laid down the weapons that could have swayed the broad public and become by definition unable to convey what it so desperately needs to convey. In the communications arms race, science is hopelessly outgunned by the juice, scandal, gossip, blood, guts and glory available 24/7 via a range of inter-connected media.

With its full-frontal assault and the limited weaponry it possesses, science arrives at the front door of perception and says that it has come to deliver a message.

The entry is fortified and the gate formidable. Not even a tank could bash through. The armed guard behind the gate, looking through the peephole, shouts something very unflattering without even opening the door or bothering to ask about the message. There is a clear suggestion that science should go far away, message and all.

Science, unperturbed, stands outside the gate and delivers the message

– trusting in a rational and even-handed reception. The guard, unseen, gesticulates obscenely to indicate his boredom and doesn't even bother to tell the owner about the ill-fated message, nor about its ill-fated messenger.

Having done its job, science leaves.

***

That was the Storytelling of Science. But what then of the Science of Storytelling? We know how scientists like to tell stories, but do we know how, scientifically, stories should be told?

In our time we might live in a world where truth and lies aren't separated by much. Where people are quick to dismiss anything that they're not certain about as a lie. But through the ages, getting listeners to give a story a chance has been a challenge.

At the very time when the Enlightenment was squeezing all the juice out of the orange and reducing it to pith, pulp and skin, the leaders of the Romantic Movement in England were wrestling with this challenge. William Wordsworth and Samuel Taylor Coleridge collaborated to produce, in the last moments of the 18th century, the anthology of poems called *Lyrical Ballads*. In trying to get readers away from the rampant rationality of the day, Coleridge asked himself how he could get them to "suspend disbelief".[96]

Which is a very descriptive term saying, in effect, "Give me a chance to enchant you. Sit down with my story, play with it like's it's the family cat or dog. Give it a chance to become your friend. See if you like it. I promise you'll like it."

What about stories, literature, music, art? Especially stories? They peddle captivation and thus beauty.

There is a primary school teacher called Ella, who is a wonderful storyteller and also a very good choir leader. Every year she leads the school's choir to win all the competitions in her region. One day, a television company calls the school saying they want to make a documentary about this extraordinary choir, and record some of their songs. It is agreed and on the given day, the television crew arrives. They start setting up. At first, the children are excited by the whole television thing, and as they wait a long time for the cameras and sound equipment and lighting to be just ready so shooting can start, they are obedient and orderly. But it takes so long, ever so long, and soon they have excess energy, they're tired of sitting still, they start becoming boisterous, then chaotic.

Ella has seen all this before. She sends them out for ten minutes and when they return, order is restored. But still the television crew isn't ready. And so Ella starts telling them a story. "One day," she says, with large eyes, gesticulation and much intonation in her voice, promising something wonderful, "there was a wicked, wicked witch with three broomsticks..."

Ella has deliberately used meter to underline the half-rhyme and alliteration in her sentence. She pauses, does it again, with emphasis:

> "A wicked, wicked witch
> with three broomsticks..."

She continues with the story that she knows so well that she never has to think about content. Just rendition for maximal effect. Hands, face, voice, eyes – she enchants. And the children stop, and listen. They stop fidgeting, stop pinching one another, stop thinking about their sandwiches. Soon you can hear a pin drop. And a minute later, as the television programme director looks at her crew to see if they would be ready to start by the time the story is over, she is amazed to see that they too, have stopped working. They're so quiet that they seem frozen. The one's jaw has dropped and his mouth is half-open.

The potion has enchanted not just the children, but the adults. Disbelief has been suspended, all resistance has crumbled.

That is the power of the story, of art, of fiction.

Science has investigated this extensively. 'Suspension of disbelief' is a well-known term now, of interest to an array of experts ranging from cognitive psychologists, evolutionary biologists and neuroscientists. It is used to explain an implicit contract or agreement between the creator of fiction and the audience. The creator's primary responsibility is to enchant the member of the audience – towards this end the creator creates a 'secondary reality' and strives to make it inherently consistent so it is most believable.[97] The leap of 'poetic faith' required of the audience may be quite small (Russian soldiers in a film speaking to each other in accented English) or may be large (Superman flying). Either way, the audience 'buys into' the fiction – that is, agrees to believe in a parallel world that is different to the one she or he lives in. In this manner she or he engages and plays with the fiction, giving the creator a chance to pull off the enchantment. This first step – suspending disbelief – is regarded as an essential element of a successful engagement between creator and audience.

When it works, the result is truly magical. We've all experienced it:

- enthralled, we no longer perceive our bodies or perceive our environ-

ment;
- somewhat gullible, we no longer judge probability or reality-test;
- surprisingly, we respond emotionally to the fiction as though it were real.[98]

Let's stop for a moment and consider what this means; With Blackbeard's Ghost, we saw that a woman and a child on a seesaw could often fail to outweigh a small toddler on the other end, all because of the defence mechanisms we have against truths that we do not want to know about.

Here, in contrast, we see fiction leaving the subject immobilized and defenceless (modern research confirms this).[99] There is always an initial period when fiction is being put to the test, when in the case of the written word the reader of a book asks herself or himself: "Shall I submit to this?"

In the case of this book, that point was reached in the very first sentence, in fact in the very first word, "Imagine we're on a yacht in the middle of the ocean, six of us."

And if you're here, in chapter 26, about to find out how all of that ends, then you suspended disbelief. You agreed to 'imagine'. You gave permission to be enchanted.

And so it is with each fictional account: the first battle is decisive.

Once that is won, the reader will entertain whatever is put forward, within the bounds of the agreement: the story has to be inherently consistent. The yacht cannot suddenly fly.

If that law is observed, at the worst of times, the seesaw suddenly balances when two evenly sized kids are perched at the ends. Reason has a chance to be weighed fairly. At the best of times, emotion is released to weigh down the side of the truth we are conveying.[100] The seesaw is unbalanced again, prejudiced towards the truth we're conveying, and it easily gets the better of what would have become the flimsy counterweight of the status quo.

Just as stories enchant, so too does art more generally. A painting, a mime artist, a poem. A scarred man on a yacht, playing his guitar and singing a song about a beggar and a businessman. He's taken off his shirt, and the evidence of the damage that occurs when a reef collides with flesh and blood is plain for all to see. The music lowers the resistance, ultimately evaporates it. The young woman who's been reading the novel looks at the singer as he improvises between the first verse and the second. It's dark, and she knows nobody will see the tears on her cheeks. She has no thoughts, just a surplus of feelings, an irresistible sense of the poignant. She thinks of her brother and her dog and the swell, getting higher. Next to her, the man and his wife are holding hands. He puts his other arm around her shoulders.

> I've got his dreams,
> he's got my freedom
> he sees my suit
> and I envy his smile ...

The wife leans the back of her head against her husband's chest and closes her eyes. The singer's voice sounds hoarse and raw, more so than she noticed before.

When a body collides with a reef, the reef wins. That truth is plain to see, he's taken off his shirt. But that truth is hard to face. It feels better to believe that the reef is far away and not a danger.

At the heavily fortified front door of perception, the gruff guard still stands ready to rebuff anything and anyone without a pass – meaning whatever does not accord with the belief system and/or happiness of the owner. Inclusive of anything suggesting a sad ending, occasioned by a violent clash with a reef.

But perception has a back door, unguarded. It only takes a credible promise of enchantment to open it. A little bit of beauty. Perhaps as little as a metaphor.

And the music plays on. Enchants. Opens the back door. And the truth about the danger of the reef, manifested in a deeply scarred body, is carried in on the song. And there, it awaits a metaphor about a truck and the Arc de Triomphe, to jointly become all-powerful and all-conquering.

And here-in lies the secret: not to join the global rush to dress up fiction as fact, but to package and sell our fact as fiction. To rejoin the truth of science with the powers of conviction of the arts. To heal the divide.

To align the people on the yacht, to address climate change, to reverse biodiversity loss, Coleridge and Wordsworth need to shake hands with Alexander Pope. Newton has to rekindle the passion in his heart and look at the stars with Walt Whitman, seeing them not just as gaseous bodies in an expanding space-time continuum, but as fixed points of rapture and solace in our emotional universe.

Socrates must sing, Nero must take the oath of truth.

A world divided between the Enlightenment and the Romantics, between science and the arts, between fact and beauty, needs to be put back together again.

By all the King's horses, and all the King's men.

27

Broaching

Erin is petrified and Julia is still extremely volatile. It is the moment in which there is the slightest window of opportunity to convince before it's too late. The opportunity must be seized with the required level of panache. A dull if urgent account will fail. So will anything too elaborate. A crisp metaphor is required and complete confidence is conveying it.

As if to remind Juan-Franco, the yacht plunges down the far side of a breaker and all three of them grab at the nearest support to keep their footing.

"...Imagine we've hitched a ride on a truck to Paris and the Champs-Élysées...", Juan-Franco begins urgently, "...The driver is speeding and we're getting there real quick. Great stuff. But when we approach we see, from every of the 11 other directions that join together at that place, eleven other very large trucks approaching, all at immense speed. We have to slow down immediately and turn away, or we're dead!"

Erin's eyes are even more frightened than before. Julia looks non-committal, unsure what to make of what he was saying.

"That's what's happening now!" Juan-Franco resumes "In the eye of the storm, tsunamis will come from every direction. We have to turn to starboard immediately, and then further, into the waves and wind. It's our only chance."

Erin looks at Julia, sees her calming down and nodding in agreement, nods too. She's paled while listening. Then Juan-Franco lays out his plan. "Please go upstairs and tell Angela you've come to relieve her. I need to tell her what's happening too. Sam's probably been lying to her and Lorenzo. Don't say anything to Sam, not yet."

She nods and leaves, her steps gingerly and scared. Two minutes later a dripping wet Angela comes down the steps. Juan-Franco saw the look in her eye just as he left the deck, and knows she's almost onside.

"Are you okay?" he asks. She nods, but she knows she's about to receive bad news. He tells the same story again.

"Sam said it would ease off," she replies, slowly and regretfully, turning at Julia, the scientist. "He's wrong...," Julia says, not losing her temper this time. He's completely wrong, and if he read what I wrote, he was lying. I don't know why."

"But you could see this for yourself when it got light, couldn't you?" Juan-Franco enquires. She nods, sadly. "What do we do?" she asks, lifting her chin.

He asks her to go back up, sit next to Lorenzo, tell him what's going on. She'll have to shout in his ear and keep it short. They'll give her five minutes and then all come up and tell Sam the majority has decided to change course. She nods and leaves. The hull is still humming, but harder than before, and the angle of ascent and descent of the yacht up the breakers is so steep that they're practically gliding up and down sky-scrapers. "Let's batten the hatches," Julia suggests, and they do so. It takes perhaps three minutes. When they're done, they decide to stow all loose articles in the saloon, except the First Aid kit, which Juan-Franco takes out from one of the holds, wraps several times in a blanket and puts in a corner. In the cabins too they secure everything except the remaining blankets. Juan-Franco lodges his guitar between the mattress and the bed so it will not be thrown about. Julia looks at the radio once more, thinking about a Mayday call.

"The chance of being heard is close to nothing," she says. He nods encouragingly nevertheless. "MAYDAY, MAYDAY, MAYDAY..." she starts soon after, adding the name of the vessel, their approximate position and situation of distress.

The static crackles back mockingly.

Juan-Franco remembers a line from Pink Floyd, *The Wall*: "Is there anybody out there...?"

"Right," he says, hugging her. She's clumsily frozen at first, then over-eager as she allows herself some reassurance. They break again, and he walks to the steps with her close behind. It's been just a day since they stood at the radio and everything was blue and perfect – including her marriage.

As he ascends, Juan-Franco thinks of the title-seeking boxer coming from the dressing room to the roar outside, the football player in the tunnel, the tennis player about to emerge on centre court. That big walk to a make or break performance on the biggest stage.

The roar is wild as he puts his head out, wilder than before and with no intention of subsiding. The other three are sitting in the cockpit, drenched. Sam, jaw set, is steering with a thousand yard stare, as if they don't exist. It's his universe, and he's the master. The spray is worse as they go through a trough, then better as they ascend. It's started to rain, and it comes almost horizontally from the stern even as they travel at speed. It's light now, and the sea is incomprehensibly chaotic and troubled. Black and blue, grey and purple, white, bubbling, seething.

Juan-Franco takes a position slightly in front and to port of the helm, about an arm's length from Sam, and waits for Julia to emerge and sit down next to Erin. Sam knows something's afoot, and braces himself. Juan-Franco stands very quietly, making him wait. Eroding his confidence, building the tension. He puts his head forward as if to speak, then bottles it again, waits.

"WHAT IS IT?", Sam finally relents, loudly and with irritation in his voice. Acknowledging him, and them. Forcing himself to listen.

"THIS ISN'T A STORM THAT WE CAN OUTRUN, SAM!!!", he shouts back. "IT'S A CYCLONE AND IT'S SWALLOWING US!! WE'VE TALKED ABOUT THIS AND WE ALL AGREE, WE HAVE TO CHANGE COURSE!"

Sam laughs, silently, incongruously, throwing his head back. His eyes flit from face to the sea to another face and another and the sea again – to see whether there truly is agreement amongst them all.

"TAKE IT!" he shouts at Juan-Franco and starts stepping away from the helm so quickly that they all freeze in shock. Then Juan-Franco, utterly surprised, recovers his composure, steps in and takes the helm – knowing a second of neglect would see a loss of all control.

They're bearing down on the next breaker, for a very short moment neither ascending nor descending. He focuses on steering very slightly to port to avoid an angle too sharp to climb when the nose meets the foot of the wave, then pulls back to starboard and dead downwind as the nose starts to rise.

Sam moves to where Juan-Franco was, closer to the bow. Juan-Franco wants to watch him, doesn't trust him. It was too easy. But all he can do is use his peripheral vision to monitor developments, and continue to steer as best he can. The other four are transfixed, knowing something's afoot. They see Sam's hand going to his chest, zipping down his oil skin, disappearing below his armpit, and reappearing holding a firearm.

He points it at them all, in turn, point-blank.

"THIS IS A STORM…!", he shouts, his voice somewhere between disdain, elation and delirium. "NOW GET DOWNSTAIRS AND LET MISTER SUSTAINABILITY STEER US TO SAFETY!!!

He points the gun high and to starboard and pulls the trigger. The shot is incredibly loud, even in the cacophony. Everyone is startled, shocked, and Juan-Franco just manages to keep concentrating on the nose and the breakers and what at that moment is a steep descent.

"GO!" Sam shouts again, "TAKE YOUR MISERABLE BEINGS AWAY FROM MY FACE!"

He pulls the trigger again, sending another bullet high through the rigging. This time they aren't frozen by shock, but jolted to action. Led by Erin, they scramble downstairs. Just seconds later Juan-Franco finds himself alone with Sam and the gun.

He steers. Sam stands, gun at the ready.

The wind is unbelievably loud – and yet the intense discomposure. The ultimate counterpoint to an uncomfortable silence. Juan-Franco decides to be quiet, not to enter Sam's world, to force him out. After all, he's the one needing a gun to underwrite his point of view.

"IT'LL PASS SOON!" Sam asserts, still screaming. It seems utterly bizarre to the point where Juan-Franco has to suppress laughter – a narcissist brandishing a firearm and shouting words of comfort and reassurance, and the only one able to hear him is the person at the wrong end of the weapon.

But then, Sam's actually talking to himself. Seeing super high waves getting higher, feeling an unbelievably strong wind, getting stronger. Somewhere inside, Sam is petrified, but for whatever reason, he has to believe that this is just the tail end of a storm. That calm waters lie close ahead and they're incredibly fortunate to have such a favourable wind to carry them there, quickly.

"YOU THINK?" Juan-Franco shouts back, to win time as much as anything else. At that volume of voice in a forty knot wind, there isn't any distinction between sarcasm and naivety, and he feels quite safe with his choice of words.

"JUST NOW!" Sam says, less loud. Juan-Franco realises that he's retreating ever further into himself, talking more to himself, becoming more dangerous.

"WANT TO HEAR A STORY?" Juan-Franco inquires, wondering how to approach such a maze of psychological complexity in such a noise, at gunpoint.

"NOT ONE YOU CAN TELL!" Sam declines, firmly.

And that is it. What more to do? Juan-Franco steers – up the swell, down the swell. Steeper, higher, with the hull snarling like some vicious, extinct, pre-historic carnivore. There's more of everything, always more;

swell, whitecaps, spray, noise, rain, wind, danger, doom. He's played every card he knows. The end is minutes away, perhaps seconds – but absolutely inevitable.

Juan-Franco jerks his head to port, casts his eyes to the area behind Sam, feigns shock. Sam falls for it, and looks over his shoulder. There's perhaps a second to act before he knows he's been conned.

Juan-Franco pulls the helm hard to starboard, lets it go completely and falls to the floor of the cockpit. He knows the yacht's so over-powered and unsteady that no further action is required. And indeed, as Sam's head is still swivelling back to where Juan-Franco was standing a moment ago, the nose of the vessel veers to the right, slightly, then exponentially more as the wind and swell gets hold of the stern, broadside. At the same time, the vessel heels violently to port and the wind gets under her, the swell too. The impetus is immense, and she continues to heel, further and further. Juan-Franco sees the water level coming at him as the yacht capsizes completely to lie on its side and the mast buries itself in the water. He holds on to whatever his fingers can find and sees Sam – overboard next to and below the yacht, hanging on with one hand hanging to the grab rail. Would she roll over completely? The keel was probably in the air, its weight fighting the combined force of the wind and swell now hitting broadside from starboard.

The yacht balances, then rolls back to a more level position. The mast snaps three feet above the deck with an immense crack, leaving the spinnaker and mainsail in the water, attached only by the halyards. A second later the next breaker hits them, and the process is repeated; heel over beyond ninety degrees, balance, roll back. This time Juan-Franco hears hysterical shrieks from below deck. The sequence would continue, on an on until they went over completely or the boat started to break up under the stress, or until it flooded completely and sank. Unless they can turn the bow into the wind.

Juan-Franco crawls to the aft hold. As he's about to open it, the yacht rolls for the third time. He reaches quickly for the grab rail at starboard with his good hand and just hangs on while the death roll goes through another episode. When the boat tips back, he quickly opens the hold, knowing that he has perhaps six seconds. He removes everything at the top in search of the rope he tied to his improvised sea anchor. He finds it too late, and though he has it in his fingers when the next wave hits, his body isn't secure yet and slides to port with the heeling, collides violently with the grab rail. There he hangs on through another entire roll to get his breath back. Below, the women's voices can still be heard, shrieking.

Through the next six rolls he crawls the thirty feet along the grab rail to the windlass hold at the bow, his good arm and leg encircling the rail, the other holding the rope. He gets slammed against the rail and the deck all the time, and the pain is immense and all consuming. Between rolls he opens the windlass hold and ties the rope securely inside, then starts to crawl back to the aft hold. He's there another four rolls later, and it takes him three to get the sea anchor onto deck. He curses his weak side, but knows that if he can get the sea anchor overboard, the rope will tighten gradually as the swell washes the yacht further away, and starts to pull on the bow, and if they're very lucky the bow will align into the wind and stop the rolling.

It's then that Sam, having come from somewhere, kicks him in the ribs, and another time before the next wave hits. As he looks up, he sees Sam holding onto the sun dodger to keep his balance through the death roll. As the boat rolls back, Sam kicks him again, and again, this time in the head, shouting indistinguishable abuse and hatred – his face bleeding and deformed with rage. Juan-Franco is on his back, powerless to defend himself. He blacks out for a moment, knowing he has to act soon. When his vision clears, the yacht is about to roll again and he sees Sam lifting his right leg again to kick – and his left leg standing half on the rope tied to the sea anchor, which Juan-Franco was holding in his weak hand when the assault started.

Juan-Franco adds his strong arm to the weak and pulls as hard as he can on the rope. He looks up as he does. The rope slips beneath Sam and his foothold starts to go. Just then the boat rolls, and Sam's eyes widen with the realization that it's all over. With no foothold and nothing to grab, he flies off the deck like a rider thrown by a horse, falls behind the sails still suspended in the water and quickly disappears.

As the yacht rolls back, Juan-Franco uses the impetus. Groaning so loud with pain he can hear himself above the wind, he turns over, grabs the sea anchor and throws it overboard to starboard, then falls down in the cockpit and waits. He's done all he can do.

Four rolls later the nose turns into the wind, and two lesser rolls later, the yacht stabilizes, finally facing the wind and the waves. She lies low, heavy with water, but the sea anchor holds.

If the other four are okay and can help him now, they might just have a chance.

28

BEAUTIFUL TRUTHS

The condemned Socrates was awake as light crept into his cell. The sun had to be rising. He thought about his plight, facing the death sentence. He'd been striving for pure reason, and it seemed like an anti-climax.

It was the fourth day of the 30 more that he would live. His food would come in a while, then Crito and the others would come to visit him and talk for hours about matters of philosophy. Which was okay. But in a way he felt that he'd thought all thoughts that mattered already and that it was enough now.

They still got something out it, Crito and the others, but the joy of playing with concepts and ideas was diminished for him. One day was like the next, and they were all so much like the days before the trial. Sure, in twenty-six days, give or take one or two, he would die. It might spur him to make the most of time, while he had it. But how? What was there to do in this while between now and the time his food arrived? What specifically should he do –something profound that would complete and enhance a life of 71 years? He'd never been into cheap thrills. And if there was nothing entirely specific to do right at this moment before breakfast, then logically there was never anything to do between that moment and the end. The little period before he would eat was a sample of all the time he had left ... and of all eternity. If nothing about it was or could be remarkable, then nothing could be remarkable about the time he has left, even if it stretched to all eternity.

Indeed, he thought again, death might just be the greatest of all human blessings.

A large ant walked across the floor of his cell. What to make of it? Fellow sufferer? It might escape easily through the many cracks in the wall. And yet it was sure to die, just like him, perhaps even before him. Ants don't live long. Perhaps then a creature blessed, like him, with the inevitability of death. But without the immortal soul.

His thoughts were interrupted by a noise outside, which at first he thought must be the guard with his food, but it turned out to be something else that he couldn't distinguish.

His thoughts returned to cracks in the wall.

There was another similarity with the ant; he, Socrates, could also escape quite easily. Sure, he managed to antagonize everyone at the trial to the extent that the hardliners won the day and he was sentenced to death – but that was four days ago, and already there would be many in Athens wishing that he'd rather been exiled – so he wouldn't be on their collective conscience.

This guard who would bring his food in a moment was not a rich man, nor a principled one, and a small bribe would be sufficient for him to leave the cell unlocked and look the other way while the gadfly of Athens walked out, fled to a neighbouring city and enjoyed his old age away from the animosities of this city. Indeed, most of the Athenians would be pleased if this came to pass, and the guard would hardly be punished.

Crito, Simmias and Cebes all had money available, and all wanted him to escape to Thesaly or another city. They said he should think of his family. But he wouldn't, for that would nullify the victory of Reason and Virtue that he'd just won. If he, Socrates, were to subvert the laws of Athens by escaping, his glorious defence of those self-same laws would be reversed. And anyway, he was tired. Tired of fighting. And in Thesaly there would be wicked people again, in addition to the virtuous ones, and it would start all over again.

Each blessing has the right time to be accepted.

And yet, that one loose end that had bothered him for all the years – the dream about cultivating and making music. He'd never done it, despite having dreamt the same dream ever so many times.

If he could learn music between that moment and the time when his food arrived, that might be time well spent. Again though, he thought about how unlike him it was. Why was he having this instruction from Apollo or Dionysus? He who had other talents? He was the one who was ruled by reason, more than anyone else in all of Athens. He wondered momentarily if there was an entire realm of knowledge and wisdom to which he'd never, ever, for one moment had access. Was the lyre perhaps the key?[101]

He didn't know, could never know now. Not in time. He couldn't learn to play even something simple on the lyre in 26 days, even if he could put his intellectual nature aside, which he couldn't see himself doing.

The guard arrived then with breakfast. To Socrates, who didn't care much for food, the event was more meaningful as a time marker in the day. It told him when he might expect his friends to come and visit him. And while the philosophy they would discuss would teach them more than it would teach him, the company was a good thing, and lit up his day.

He put the food aside for another moment, trying to order his thoughts about the dream, and the music. At least, he had to do something, as it seemed the gods were ordering him to – and he had to obey.

Perhaps he could compose something. That might not qualify as making music, necessarily – but then it might. At least he'd be cultivating it.

He would do something in the 26 days.

***

In the centuries leading up to the life and times of Socrates, the lyre was the ultimate instrument for expressing emotion. Originally created by Hermes for Apollo, it was used to accompany the performance of lyric poetry, where all the heart's strongest yearnings could be expressed. It could have been any latter day rock star slowing down the pace after an energetic number and singing a ballad, guitar in hand.

Indeed, in the four centuries between Socrates and Nero, the cithara had become popular as the instrument of choice of the professional musician. It still looked very much like the lyre, but the two strings had been expanded to eight. It had a formidable sounding box and was thus very much the big brother of the lyre, and yet undoubtedly of the same family.

And yes, it was still about feeling and emotion. It still accompanied lyric poetry, epic recitations, odes and rhapsodies.

The lyre that Socrates was still spurning in those last days of his life would become the cithara that Nero would play wherever he went – perhaps even as Rome burnt. It was also spelt 'kithara'. And some two millennia later, adapted in design and sometimes electrified, it would become the 'guitar' that Jimi Hendrix would set alight while playing the *Star Spangled Banner* and that Bob Dylan would play as he trawled Greenwich Village as a young man, from gig to gig. The guitar that Juan-Franco would build from a book when he was 12. The instrument he played at every one of his shows and that last he played as the sun was setting over the yacht, the evening before they sailed into the storm.

The lyre had been calling Socrates for decades, saying; "Come, play me.

Understand that there is a world beyond the logical – another universe. An alternative form of persuasion. Come, let your truth be carried on my notes."

But like he always did, Socrates had disagreed and declined. Indeed, he'd veered even further the other way, towards perfect rationality. And now, approaching death and craving the release it would bring, he'd still not made music. No, his last days were spent fortifying the door between the universes. Doing just enough to appease Apollo.

And as he finally passed, the lyre would still wait to be played by him. His legacy would be a tradition, a metaphysics that would influence countless generations of scientists, stoically refusing to rage against the dying of the light, even if it should be the light over all of us. That would affect also the esteemed authors of the Intergovernmental Panel on Climate Change in writing their reports. And would influence Julia, walking onto the deck of the yacht in the time before the mast snapped. Causing her to feed her audience only cold and hard facts.

Ugly truth was the legacy of Socrates. And deprived of so much that matters, how could it not be ugly?

***

Julia was trying to warn of an impending storm.

Elsewhere, a new serial killer was discovered. It's named PECCA and is responsible for a variety of deaths, ranging from the effects of climate change to those emanating from biodiversity loss to the fall-out from over-consumption/over-population. Many people, including scientists and environmentalists, see the damage these pending storms will do and have been trying to warn of them.

Both categories of people take their master model from Socrates, and both are imperfect.

The scientists publish and disseminate ugly truth, just like Socrates did. They talk about what they see happening, mostly to other scientists who only make up a small fraction of the population, whether it be of Athens or of our modern, global village. Even when they talk about policy matters ultimately decided by voters rather than panels of scientists, their approach stays exactly the same.

They stick to the truth, the whole truth and nothing but the truth. This makes them happy. They feel elevated above people who package the truth. Speech writers, and newspaper editors, and novelists.

And yet, they are the sentry in the tower, the look-out in the crow's nest, the little shepherd watching out for a wolf. Their job is to warn us when a menace is heading our way. Not just to see it but to tell us. And not just to tell us, but to tell us in a way that we can understand and that galvanizes us to take the necessary action.

If they communicate to us an ugly truth and our attention is diverted by a beautiful lie so that we don't hear, then they have failed in their duty. Following Socrates and the tradition he started, they fail from time to time. Not that often. But often when it matters the most.

Then there are the environmentalists. Most of Juan-Franco's friends are environmentalists, even though the majority of them do things very differently to the way he does them. They mean so well, their hearts are in the right place. They see menaces coming our way and worry about their impact not just on themselves and their family but on all of humanity and all the species on earth. Their view is the one from afar. They see the story of the whole, and they rebel against the dog-eats-dog reality of the story of the parts. To them, if it's clear that there will be water shortages in future, it's incomprehensible to respond by buying shares in a desalination company and revelling in the money they are going to make. That is for the Sams of this world.

And yet, in their interaction, are many not just like Socrates? The man who was the conscience of his society and who had more insight and intelligence than his fellows? And who abdicated his responsibility to lead them on the right path by deliberately leading them astray – provoking them to kill that very conscience? This at best would end in a societal lobotomy, and at worst in a societal suicide. But Socrates was willing to sacrifice the future of Athens in order to win a victory for Reason and Virtue. And indeed, only a few decades after his death, Athens lost its political power for good.

All along, Socrates was one who criticized from the outside; "I am virtuous and reasonable, you are capricious and barbaric. You are pretty average, I am somehow superior to you (just ask the Oracle of Delphi). You killed the eight generals, not me. You killed Leon of Salamis – not me".

And yet – if he was the conscience of Athens and the wise counsel, these abominations were as much his failure as theirs. The society as a whole failed – and in this, Socrates failed with them.

On a yacht facing a cyclone, there is little point in having been right about what should be done if the yacht is pulled asunder and everyone perishes. Shouting "I told you so!" just before the bow disappears below the stormy waters is cold comfort. The yacht must survive. The whole, the society, the civilization must survive.

Many environmentalists today are following the example of Socrates. They criticize from the outside; "You fly too much, you pollute too much, you eat too much red meat, you're having too many children. You're deforesting, not me. You're causing the extinction of countless species, not me. It is you putting narrow self-interest and profit over the planet, not me. You are spoiling everything for our children, not me."

Just as it was with Socrates and Athens, the role of these environmentalists is to ensure that the yacht survives. If they cannot manage this, it will be a collective failure. And there will be cold comfort in being proven to have been right, when it's too late for anything meaningful to be salvaged.

***

To the end, Apollo wouldn't leave Socrates alone.

He was convicted right at the start of the summer. It was the feast of Delian – 30 days devoted to Apollo. A sacred boat would be dispatched on a pilgrimage to the island of Delos, near Mykonos, where the god was born. During its absence, no criminal could be executed.

And so Socrates lived a while longer at the mercy of Apollo and the weather, for when the ship returned, his time would be up. Being at the mercy of Apollo he felt he had to do something to satisfy the command he'd received so many times in the dream about music. And he complied. He didn't do a whole lot, but a least he did something.

Then Crito came early one morning, saying that the ship was rumoured to be close. He urged Socrates to escape. They talked for a long time, debating this, and Socrates ended up winning, as he always did. He would not escape – it wasn't the right thing to do.

The ship returned soon afterwards and all knew it was all nearly over. That day stretched and stretched – it was almost mid-summer now and the white heat that lay over the flat parts of Athens between the hills found its way into the cell. Then, the day was finally done and all that remained was one more night. It became cooler, and quieter.

On the last morning, Socrates was taken out of his chains. His wife Xanthippe was with him, holding their young child, very distressed. Socrates asked that someone take her home. Emotion departed the scene and left rationality behind. After some reflection by Socrates on pleasure and pain, Cebes asked on behalf of Evenus the poet,

"Socrates, you never cared for poetry, why are you now putting Aesop into verse and composing a hymn in honour of Apollo?"

Socrates, smiling slightly, replied: "Tell Evenus that I know I can't match him or his poems – but I needed to purge away a scruple. In the course of my life I have dreamt that I should make music. The same dream came to me sometimes in one form, and sometimes in another, but always saying the same or nearly the same words, "Make and cultivate music", said the dream... I thought that I should be safer if I satisfied the scruple, and, in obedience to the dream, composed a few verses before I departed. And first I made a hymn in honour of the god of the festival, and then considering that a poet, if he is really to be a poet or maker, should not only put words together but make stories, and as I have no invention, I took some fables of Esop, which I had ready at hand and knew, and turned them into verse."[102]

He could not make stories, and had no invention, and so turned to an existing story and made it into a hymn.

His end was to be a far greater blessing than he'd ever imagined. The poison hemlock was chosen as the method of his execution. It was a benign death, more so than any other method would have been. Moreover it extinguished all feeling. His legs would go numb, then his thighs, perhaps next his fingers, until he felt nothing at all. And all the while he'd be in complete command of his senses, talking to his friends about philosophy and about this world and the next and about why the outcome that was unfolding was the right one and not to be grieved over at all.

And so it all happened. Harmony was discussed at length, and the lyre came up as a metaphor. The guard told Socrates to walk around, quickening his pulse so the hemlock would work faster. A while later, the philosopher lay down. The guard came and pinched his foot – he could feel nothing and the numbness would now spread throughout his body. It crept up, up, towards his heart. He couldn't feel anything. Just nothing. But he kept uttering intelligent statements, and conversing. Remembering about a debt that was still due to Asclepius, Socrates urged Crito to pay it back. And as Crito confirmed that it would be done, Socrates' head fell to the side, his jaw dropped and his eyes went glassy.

It had been truth, always the truth, nothing but the truth.

The one who never cared for beauty was gone.

He'd achieved at the very end a total lack of feeling, and a cold rationality. Like Ernest Rutherford would split the atom, Socrates had split off rationality from the body, logic from emotion and the individual from the whole.

Man isolated from nature would follow in due course.[103]

***

Julia stormed out with a warning about a cyclone. When the others wanted to know more, she reverted to type and to habit; she started to explain like a scientist does. She returned quickly to ugly truth. Even the document she wrote by hand and tied to the mast in a plastic sleeve contained nothing that was beautiful. Unsurprisingly, it was thrown in the sea before ever being read.

Sam had his own agenda. He peddled a simple and beautiful lie; "Everything will be alright". The beauty of it lies in an implied higher order of things and in the conclusion that because of this higher order, nothing bad would happen. Lorenzo's grandfather had the same mindset, but in another way. Beautiful it is, but the victims of Auschwitz or of Stalin's Gulag or of Pearl Harbour or of 9/11, know it to be a lie.

Still, for a good while, as the yacht drifted on the current and was pushed by the wind into ever greater danger, this beautiful lie prevailed, because all that opposed it was Julia's ugly truth. And the beautiful lie is stronger than the ugly truth.

Juan-Franco was looking for a way to communicate beautiful truth. Something that was unusual, or skilled, or impossible, or that was dangerous, or ground-breaking, or appalling. He thought about something that would transfix. He found an opportunity just before the sunset, and took off his shirt to reveal his broken and scarred body. He appeared to be there only to play guitar and sing and the image of what a body looks like that has collided with a reef had the chance to slip in the back door of perception – carried by a tune. The music was the Trojan Horse that obscured the true message from the guard at the front gate.

And once it had entered the back door, it lay waiting to germinate.

The waves grew – the wind got stronger. But Sam found a way to employ propaganda and Blackbeard's Ghost – the prejudice we all have for good news. The imbalance that causes the rational seesaw that weighs evidence to come down on the side of the lighter weight.

"It's just a storm," he asserted confidently, having just thrown the scientific evidence into the sea. And as he was believed and followed, for another period of time they moved ever closer to the eye of the storm, this time under full sail.

At this point, Juan-Franco found a metaphor to describe the fact that they were approaching disaster rather than salvation. He told the others about a speeding truck approaching the Champs Elysees and seeing eleven other speeding trucks approaching from every direction. His metaphor conveyed that they weren't at the back end of a storm, but in the perimeter of a cyclone. And interpreting the rising waves and strengthening wind in this context, it was clear to everyone.

Except Sam. To him, there could be no truth more beautiful than the one where he gets to the island in time to restore his financial empire, retain his standing in society, and safeguard his self-respect. For Sam, no evidence, no story, nothing could ever be stronger than this.

This is not so unusual, for the world has a dual DNA. There is collaboration and competition. Sam could not be brought around to look past competition. He'd placed himself directly and irremediably between the salvation of the yacht and the consensus of all the other crew members who wanted to move there. In this manner, he ensured that the impasse would be resolved by conflict, for the time to talk had run out.

And so it is with PECCA's, the new serial killer on the block. In biodiversity loss and climate change and other PECCA's, beautiful truths may not produce complete consensus, but they may very well bring a sufficient agreement to change course – dramatically if need be. And even if vehemently opposed by some – if need be.

Juan-Franco in his small way has managed a modest and localised reconciliation of the world Socrates once split apart. Usually, he does so through his musical performances and his small business ventures in sustainability. Lately, he used this approach to bring around some doubtful people on a yacht, trying to avoid a cyclone.

Surely others in other places and other fields also bring together things that belong together. Like beauty and truth.

Taking his modest synthesis further, were Juan-Franco ever to write a book, he would probably proceed by changing the identity of the protagonist sufficiently to make himself unrecognisable – and above all by constantly reminding himself to tell a story.

29

The yacht

The helicopter pilot looks at his instruments and again shakes his head in disbelief at the enormous, circular mass of cloud that demarcates the cyclone. It's weakened somewhat in the past 24 hours, but it's still moving west northwest at about seven knots an hour. It will make landfall sometime tomorrow, the weather office is hoping it will be down to a Category 3 by then, with no further loss of life, and only minor destruction of property.

He's flying east of the cyclone's edge, up and down, searching. Next to him sits a navigator who doubles up as a paramedic. On board is an array of medical equipment. The people they search for are always very high-worth individuals with life cover policies exceeding US$20 million – in case of a mortality event. Free search and rescue is included in the policy as a costless, extra benefit – designed more to benefit the company than the insured. Rather spend US$100K flying over a vast expanse of water, looking for a needle in a haystack – for if the insured can be saved, the mortality event is avoided. It makes policy holders sleep better, and shareholders too.

The call from Sam Preston's office came in about 36 hours before. It wasn't easy – an amphibious helicopter was needed, and with the devastation in the broader region, nothing was available for charter. Ultimately, they flew it out of Miami after first travelling there themselves by private charter in a Learjet. Refuelling was complex and night fell the previous day before they could start searching.

The area was vast, and they had a decision to make: Where to look? If they looked well east of the cyclone, they'd be looking for a yacht that had never been in serious danger and therefore probably didn't need help. So they decided to fly up and down the edge of the cyclone, waiting to see if something popped out beneath the cloud cover. That would be a yacht hit by the edge of the cyclone – one that had somehow managed to hold its position while the cyclone moved away, so that ultimately it would enter calmer waters again.

The amphibian carries extra fuel tanks and consequently has an enormous range, but they can't go that much further before returning to the nearest land for a refuel.

"What do you think? Shall we bail?" the pilot asks his assistant.

"Give it 20 minutes," comes the reply. There are bonuses for successful missions. "Go a bit closer to the edge, let's double back, there must be another 10 sea miles that have opened up while we were flying this line ..."

They turn west, fly right to the edge of the cyclone, then start on a slow arc along its edge, heading more or less south. They could quite safely head in a bit, below the clouds, but visibility would be so poor that it wouldn't be worthwhile.

They see something 12 minutes later, go closer, notice that it is indeed a yacht. Or once was. The pilot whistles through his teeth in surprise. There's a lifeless figure in the cockpit, and another sprawled near the bow. Both seem to be attached to safety lines, neither seems to be conscious.

"That's her!" exclaims the navigator, checking the de-masted, low lying vessel's description against the one they got from the charter company she belongs to. "Can you put this lady down? Swell looks like about seven feet."

They both know he can easily do so in the conditions that now prevail, and they prepare to land.

***

Juan-Franco ascends slowly from a pit of utter oblivion. As if being hoisted from the darkness of a very deep well. For a while nothing changes, except for the realisation that there is movement and that at some point, things will change. There will be cognition, and light.

He hears the sound of an engine – not a ship's engine. Perhaps a helicopter. Is he being hoisted by helicopter? From what pit? Why does he think that it isn't a ship's engine? What do ships have to do with it?

Sensations break through the stupor; a dry mouth, a swollen tongue, something like wool on it. Pain everywhere – but intense pain below his sternum, to his left side.

He's nearing the top end of the well and becomes aware of more; he's lying with his lower body on something flat and his upper body perched against something semi-hard that pushes against his face. It must have been a while, for he is intensely uncomfortable and his back is threatening to cramp.

The engine is louder now, and it seems his hair and wet shirt are being ruffled and cooled, like they've been so much just recently. But by what?

He opens one eye – the other won't open. The engine stops, quiet descends. Except for water against some hard object.

Then he knows and remembers with a start – everything. A montage of surreal images flash before him; Enormous swells, a sea and air that are inseparable and indistinguishable, horizontal rain coming at eighty miles an hour. Blackness, blueness, grey and white, shafts of light. Incessant noise throughout the spectrum; a low humming hull, high, terrified shrieks; the crack of a snapping mast; Lorenzo's bellow of pain as his collar-bone snaps, followed by a whimper as he tries to control himself.

Commands, warnings, cries of help, all at maximum volume above the fiercest of winds. Cutting the halyards so the sails in the water would stop dragging them sideways. Watching the mast drift away and disappear.

Himself at the helm, running the engine at a knot or two to make sure they're not dragged back too much. First Lorenzo at the bow, one hand on the rope holding the sea anchor, the other pointing down or up for Juan-Franco to accelerate or decelerate – always trying to keep just a little bit of tension in the rope. Then that extra large breaker, Lorenzo slipping and falling, the cry of pain. Julia coming up to take his place, Lorenzo down to Erin and Angela, making sure the bilge pump's going flat out. Occasionally, as the waves break over them, passing along buckets of sea water to empty over the side because they're afraid the bilge pump is losing the battle.

Nose up, balance at the top, nose down, into the trough, nose up. On and on and on and on. Every six seconds, by nightfall every seven. All the time thinking about how flimsy his sea anchor is and worrying the rope would go. Especially after dark, when they aren't able to see where to go and might easily make a mistake without the sea anchor and broach again – for the last time. It becomes harder to see Julia's signs.

But the roar of the wind seems to lessen a bit – if it isn't his hearing going after the relentless onslaught. And yes, the rain stops. Starts again but also stops again. Somewhere near midnight he can't see Julia anymore and doesn't know if she's lying down or has slipped inside without him noticing. With him perhaps blacking out.

Nose up, nose down. Nose up, balance at the top, nose down, into the trough. Counting, and thinking it is eight – but not knowing. And ultimately, the engine dying as the fuel runs out, simultaneously the bilge pump stopping, and there is nothing more to do, except to hope that the cyclone will continue to slip away from them. Sliding to the bottom of the cockpit, and oblivion.

He's too tired to react to the images he recalls. There's no energy to shudder – just to remember.

The helicopter engine has died down now, and is followed by an outboard engine and voices as that engine too, dies down.

He feels the boat moving as someone ascends from the side. He tries to sit up, manages to straighten slightly.

Then a sun-glassed and capped face comes into view. A man, tanned. "Good day," he says with an American accent, "Executive Search and Rescue. Jonathan Jackson is the name... don't move, you look a bit the worse for wear."

Another man appears, and goes to the bow, where he talks to someone else. He ends the sentence with an audible "...madam." It has to be Julia then, and she has to be somewhat okay.

"We'll attend to you now," says Jackson, moving to the steps leading down to the saloon.

"Juan-Franco," he tells Jackson, and is amazed at his voice. How hoarse it is.

"Be back in a moment, Juan-Franco," Jackson acknowledges the introduction, and disappears downstairs.

There are more voices there. Different tones, high and low. The sun is out, Juan-Franco realises. The sun is shining, and the swell is down. They had to be out of the cyclone.

It's finally gone.

Jackson reappears with a drip then.

"Let's get you onto this seat so I can have a look," he says. He helps Juan-Franco up. The pain is intense and ubiquitous. He lies down again, and Jackson checks for broken bones with quick yet careful fingers. As he lifts Juan-Franco's shirt, he notices the extensive bruising on his weak side, where he'd been banged against the deck, then kicked. "Go really slowly, but breathe deeply...," he says. Juan-Franco doesn't get far before wincing and terminating the effort.

"Some pre-existing injuries and muscle atrophy, obviously", Jackson notes somewhat prosaically. "I suspect broken ribs. Lots of abrasions and bruising all over. A black eye. Dehydration and perhaps shock. Let's start with this, I'll add some pain killer..."

Jackson inserts the butterfly needle into Juan-Franco's good arm and attaches the drip to the last vestiges of the frame of the sun dodger. The canvassing is completely gone.

"How did you know about us?" Juan-Franco asks, his voice barely better than before.

"A certain Esthia called Mr Preston's office," Jackson begins, then stops. "Mr Preston holds a life insurance policy with us. It includes search and rescue. Would you be Mr Preston?"

As Juan-Franco shakes his head, he can see that Jackson suspected as much. "He went overboard when we broached and the boat suddenly heeled," Juan-Franco says. "He was standing right next to me. I saw him fly over and into the water. He was gone a second later."

Jackson takes out a mobile phone and asks permission to record the rest of the conversation. Juan-Franco nods.

"So you saw Mr Preston going overboard?"

"Yes."

"What caused it?"

"The rising deck as the boat heeled, propelled him into the sea. The others were downstairs."

"Would you say it was involuntary?"

"Absolutely."

"Was he wearing a life jacket?"

"I'm not sure, I think not."

"And do you assume him to be deceased?"

"Absolutely."

Jackson puts his mobile phone away. "Well, I understand that lady on the deck is his wife, the policy also includes search and rescue for next of kin."

He's dropped his voice, belatedly, as if to respect Julia, who's being attended to by Jackson's assistant.

"Is she okay?"

Jackson nods.

"And the others?"

"Very, very shaken and hurt. Some are dehydrated from constant vomiting. Shocked, but okay. One badly broken humerus. By the way, what caused the yacht to broach?"

"We thought it was time to change course," Juan-Franco says, his voice sounding a little better now. He takes a sip of re-hydration fluid that Jackson offers him. "But the sea was rough, and when we turned we broached and the mast was submerged. You get the idea."

Lorenzo comes up the stairs, pale, with his arm in a sling. Angela and Erin follow. They squint in the bright sunshine. Jackson's assistant helps Julia to her feet and puts her arm around his neck. They walk to the cockpit and sit her down.

"Our deepest commiserations, madam," says Jackson in the smooth manner of many other bereavements, "Your husband is presumed dead."

Juan-Franco reaches out with his weak arm – the one without the drip – and puts it on her shoulder.

The sun drops a little lower, the swell toys with the hull as if it were just another pleasant day. There isn't much to say, and nothing more is said.

"We have space for two more in the chopper," Jackson ultimately breaks the silence. "The owner of the yacht or his insurance company can decide whether to try and salvage her. We can return at first light for another two and then soon after for the last person. We can sort out the costs later."

Juan-Franco looks around. No mast but for the stump. No sails. The sun dodger gone. The grab rail twisted in several places as if chewed then spat out by a giant, angry animal. But the helm seemed to be intact, and the engines were running before the fuel ran out. And the bilge pump.

"How far to an easterly port?" asks Juan-Franco.

"About 60 sea miles I'd say."

"If you could spare me some fuel I'd prefer to motor her there," he says. "We'll be there by sunrise."

Jackson lifts an eyebrow in surprise and waits for the others to respond. "I'll stay too," says Julia. "Angela, Lorenzo is in pain, why don't you two catch a ride?"

But they all look around at the battered vessel that has served them through the worst, and the people who went with them through the worst, and remember just how close to the precipice they were.

One by one they all elect to stay. They could take it easy, be close by morning and motor past the reefs and into port in daylight.

Jackson gives Lorenzo an injection for the pain, checks their navigation equipment, makes two trips back to the chopper to fetch fuel, fills their tank and leaves a surplus jerry can alongside. He starts the engine in a test, then switches it off again.

"Will you call Esthia, please?", Juan-Franco asks as they prepare to leave. "Just tell her all five of us are okay and that we're missing her."

He stops, starts again. "Tell her Juan-Franco says hi and thanks ever so much, and that I miss her. I'll be in touch about the trip to Samos."

Jackson pauses, looks at him quizzically to understand the profundity of the moment that's just passed. "You want me to say it like that, word for word?"

He gets a nod that is more affirmative than any number of words, scribbles in a note pad. And then the two men say goodbye and that they will check in the morning to see that the yacht's safely arrived. The helicopter starts, whips up spray, takes off and disappears.

It's late afternoon, forty-eight hours after it all began.

They cut the rope to the trusty sea anchor, start the engine, eat rusks that are still dry, drink not Mochitos but rehydration fluid. The bilge pump gets back to work and the abundant fluid in the hull slowly starts diminishing.

Juan-Franco heads east. Even as they head further away from the cloud mass behind them, the sun disappears behind it and the dusk is inordinately long.

As he holds the helm, his left is touching it, feeling it. Like always it does.

***

By the time the light starts to fade more noticeably, the yacht is significantly higher in the water and the swell has dropped further, to about four feet. They've been picking up speed.

Juan-Franco cuts the engine and the yacht glides to a halt. The other four are close – two on the cockpit pillows, two more on the deck. They've been sitting quietly, saying nothing – but all in tune.

"We have to bury him," Juan-Franco says, simply. It's evidently true, and hard to argue with. "It's not like..." Julia starts, then gets stuck.

"Not like he covered himself in glory towards the end," Lorenzo helps her out.

Not one of them knows what happened yet, and no-one has asked.

"I pulled the helm to starboard when he wasn't watching," Juan Franco starts. "I knew we'd broach and I knew it would be nasty. But we were running out of time and he wouldn't hear me. He attacked me, kicked me. I defended, the boat rolled, and he flew overboard. It was over in a moment. I didn't tell Jackson what he did before that, because it makes no difference now."

With the end clear in everyone's mind, the imperative returned to pay Sam a last respect. "I'm hardly at mass or confession every second day," says Julia. "I don't think I've been there this millennium."

"We just have to bury him. You tell us how to do it," Juan-Franco says.

"I agree," says Erin. "You decide, Julia. And we'll follow you. You can recite the periodic table if you wish."

It slipped out before she could think it over. Embarrassed, she giggles, and Angela sniggers, and they all laugh for a good 30 seconds as the tension and stress roll off them. Juan-Franco, in agony with his cracked ribs, tries his best not to laugh along but can't quite help himself. When they stop, the mood is immediately serious again.

"Your sermons have been better than mine just lately," Julia says, looking at Juan-Franco. He nods, and asks Erin to fetch his guitar. When she appears and he takes it out of the hard protective case, it's mostly dry, amazingly. He tunes it as the light begins to fade. "Does someone have a dollar bill and dry matches? And a plate?" he asks. Again Erin obliges. While she's away they all move to the bow, and sit down on the deck.

He puts the dollar bill in the plate and lights it. Far beyond the cyclone and its clouds, the day is well on its way to more westerly places. As the note burns, more slowly than expected, it's bright in the gloom. Then it drops in luminosity, only a last quarter burns, and it flickers ... and dies.

Juan-Franco starts strumming his guitar, slowly and simply, and rises. His swollen and bruised face is no longer visible, just the silhouette of his asymmetrical body. Then he starts to sing the well-known hymn, just the first and the last verse, his husky voice far more hoarse than usual:

> Abide with me, fast falls the eventide
> The darkness deepens, Lord with me abide.
> When other helpers fail and comforts flee,
> Help of the helpless, O abide with me.

It's then that they fully comprehend how close it's really been, how nearly they were gone, how precious this moment and every other.

Lorenzo remembers his grandfather and the God who used to be in his heaven while all was well with the world. The one he prayed to two nights ago. And thinks; abide with me.

Abide with me. So thinks the thrashed and innocent child, sobbing behind the sofa as the father leaves him to repent for having an opinion, and to lick his wounds.

Angela looks at Lorenzo. Juan-Franco thinks about Esthia and Julia about the swallow below the window-sill and her mother. Erin thinks about her brother and her dog.

> Hold Thou Thy cross before my closing eyes
> Shine through the gloom and point me to the skies.

Abide with me. Nero thinks about Poppaea and Petronius, and his mother

Agrippina, and all his other loved ones that he killed in his insanity, to be left alone at the end waiting for the horsemen to come. All alone.

And the same words are in the mind of Socrates as he looks at the people of Athens, and in their minds as they look at him.

Abide with me.

> Heaven's morning breaks, and earth's vain shadows flee;
> In life, in death, O Lord, abide with me.

Abide with me, says art to science, and science to art, and truth to beauty. The black daisy to the white, and humanity to hope.

Abide with me.

And then the song is over.

Silence falls, profundity descends.

Juan-Franco takes the plate from Erin, and crushes the remains of the dollar note, then extends his arm over the water and scatters it all.

"Dust to dust, ashes to ashes", he starts. "Water to water, cyclone to cyclone."

For a few seconds longer he stands poised – arm extended, motionless and quiet. Regal. And in that moment, Sam is remembered, with much else.

Juan-Franco sits down then, and says no more. Prospective small talk seems inappropriate, and they all just sit and later lie on the deck.

The swell is almost negligible now. Some of them close their eyes, feel the gentle motion of the boat – almost like being rocked to a lullaby. There is a kind of oneness, a benevolent harmony, gently enveloping them.

Moments pass, then Juan-Franco walks back to the helm, starts the engine, turns east at a leisurely speed. Looking over his left shoulder, he sees the North Star, and smiles.

He looks further north, and then all the way to the south, across the immense multitude of stars, bright and reassuring in the way they're ordered – into the same constellations as ever.

On this sea calmed, the yacht heads into the night.

POSTSCRIPT

to the second edition

December 2015 – The road beyond Paris

In the world of the PECCA that inspired the metaphor of the yacht and the approaching storm that guides *Nero's Violin*, three weeks ago in Paris, Socrates rolled his boulder closer to the top of the hill than ever before. The Paris Agreement joined the 195 countries and the European Union, all of whom were represented at the 21st Conference of Parties to the United Nations Framework Convention on Climate Change, in adopting an ambitious climate plan by consensus. Each country will pledge a certain, quantifiable response and the aggregate will be reviewed from time to time to see if enough is being achieved. There will be a strong focus on renewable energy and a concerted intent to terminate the addition of carbon dioxide into the atmosphere entirely at some point in time, probably about 35 to 50 years from now. For the first time, everyone appears to be in accord.

To many, the work of two decades has finally borne fruit. They feel that the nations of the world have come to an agreement and we'll all be okay. That climate change as the archetypal PECCA is finally under control. To those concerned more broadly about the state of the environment and its people, animals and plants, this would also tend to give hope that other PECCA's can be addressed by some combination of lobbying, gentle persuasion, consensus building and direct action.

Indeed, it was a great victory belonging to thousands of people across the globe. Just the subset of those that I am fortunate to know includes a large number of utterly remarkable people.

The task of a climate practitioner can often seem daunting in the extreme. Everything to lose and so many moving parts to co-ordinate. All the human defence mechanisms of denial, rationalisation, and others to navigate every day. All the trappings of human fallibility and hubris getting in the way. And for those with their finger on the pulse of the natural world, something akin to working in a terminal cancer ward – for years the news just got a little worse every day.

Therefore, there is understandable elation at this unexpected win. And yet, the agreement does little more than state a collective ambition. The pledges of the signatories, if aggregated, leaves us with at least 0,7 degrees of warming above the quantum in the agreement (and what science considers to be safe) – perhaps as much as 2,2. Thus, the pledges do not add up to the ambition as yet.

This is due largely to a conviction gap. In most countries, politicians are faced with the reality of constituents who either don't believe in the threat at all, who under-estimate it and/or who actively lobby against any remedial action – due to narrow self-interest. There are still billions of dollars of fossil fuel subsidies in place internationally. The 'polluter pays' principle is not applied to the damage caused by the resultant emissions. Free trade agreements hamper the ability of countries to prioritise climate responses. The aviation industry does a great deal of harm but is largely unregulated. Much work is left to do, and persuasion is a central theme.

In this manner, the dynamic of the yacht plays out. And yet, agreeing on an ambition is a fantastic landmark. It means that the stage largely moves from international to national. In each country, as a start, the state should be held accountable to the pledges it has made, even if they are insufficient. And then, beyond that, we have to convince enough people so that we can move beyond, to better pledges. Already the technological interventions required often come at zero cost.

The heart of the audience has not been won just yet. When that happens, everything else will fall into place and our storytelling will have won the day.

NERO'S VIOLIN: A COMIC BOOK EXECUTIVE SUMMARY
SCIENTIFIC PEOPLE LIKE TO BE VERY PRECISE AND STICK TO COLD AND HARD FACTS.
Annual Carbon Emissions by Region
USA & Canada
Western Europe
Communist East Asia
Eastern Europe & Former Soviet States
India & Southeast Asia
Australia, Japan, Pacific Ocean States
Central & South America
Middle East
Africa
Million Metric Tons of Carbon / Year
1800 1600 1400 1200 1000 800 600 400 200
1800 1850 1900 1950 2000
EVEN WHEN THEY TELL US HOW TO SAVE OUR WORLD.
CLIMATE CHANGE
BIODIVERSITY
THEY AND THE ACTIVISTS THINK THEIR JOB IS TO PULL THE ISSUE INTO TRUTH - ONE AXIS ONLY.
TRUTH
LIES
TRUTH
CYCLONE COMING
WEIGHING EVIDENCE
THE HEAVIER DOESN'T ALWAYS GO DOWN.
HUMAN NATURE MAKES US AVOID FACTS THAT MAKE US FEEL ANXIOUS OR AFRAID.
TRUTH
WE LOVE CAPTIVATION, NOT JUST TRUTH. WE'LL PAY TO WATCH A LIE, AS LONG AS IT'S BEAUTIFUL (CAPTIVATING).
ADMIT ONE
Lies
Beauty
Lie
Truth
Ugliness
CONVICTION TAKES PLACE ON TWO PLANES: TRUTH/FALSEHOOD AND BEAUTY/UGLINESS.
STORIES AND METAPHOR GET INTO THE BACK DOOR OF PERCEPTION, WHERE WE DON'T DEFEND AGAINST THE TRUTH.
STORIES WELCOME
SCIENCE SHOULD TELL STORIES.
FOR CLIMATE CHANGE, TRY...
A COMING STORM AND 6 PEOPLE ON A YACHT.

ENDNOTES

INTRODUCTION

The story of a beautiful lie: Nero and his violin

1. Today, when we say someone is "fiddling while Rome burns", we imply in the strongest possible sense his/her complete lack of perspective; indulgence of subjective fancies while an objective crisis is playing out; and his/her total lack of strategic leadership. We want to say that to a calamitous degree the person doesn't "get it" and is utterly divorced from reality – caught up in their own fantastic little cocoon. On a sliding scale of monarchial insensitivity, it even outranks Marie Antoinette's "If they don't have bread, give them cake" (a statement which interestingly is also an accepted truth without any evidence that it was ever uttered). The person we accuse of "fiddling while Rome burns" is a basket case, utterly and irrevocably beyond redemption. On a first level, by making the title of this book *Nero's Violin*, the suggestion is that as a society, we're fiddling while Rome burns. That we've lost touch with nature; we're failing to read the most blatant warning signs or to heed them; that we're caught up in subjective matters that we consider very important while an objective tragedy is occurring before our very eyes.

2. In a fascinating study by Mary Francis Gyles, "Nero Fiddled while Rome Burnt", *Classical Journal*, Vol. 42, No. 4, 1947, pp 211–217, the author investigates the veracity of the received wisdom about Nero, his violin and the great fire of Rome. She quotes the historian Tacitus who writes of Nero: "... at the very moment when Rome was aflame, he mounted his private stage, and, typifying the ills of the present by the calamities of the past, sang the Destruction of Troy." Gyles also refers to

Suetonius who states that Nero ... "viewing the conflagration from the tower of Maecenas ... sang the whole of the Sack of Ilion in his regular stage costume".

At the time when these historians wrote, there is no mention of any musical instrument in Nero's theatrical behaviour. Just that he sings. Only a hundred and fifty years later is there an implicit reference to a musical instrument by Dio Cassius, who says that Nero sang while wearing the costume of a *cithara* player. It seems certain Nero was indeed a cithara player – it was one of the most prominent musical instruments of the time. It was widely used in the Greek tradition to accompany poetry and singing and would have been the instrument Nero used when taking part in musical competitions. When comparing the cithara to modern instruments, it resembles the harp.

The violin would not be invented for another 1,500 years and there is no recorded connection between Nero and the early precursor of the violin, the fidicula. Thus, it seems likely that Nero sang while Rome burnt. It is possible he accompanied himself with a stringed instrument. If he did, it was almost certainly a cithara. What we know with utter certainty is that Nero did not play the violin while Rome burnt.

Mary Gyles traces the development of the misconception through the middle ages to Shakespeare and concludes: "From the time of Shakespeare and the unknown author of *The Tragedy of Nero*, the trend is definitely established. Nero has become a fiddler, and in spite of the fact that every careful historian has refuted the statement, so he remains today".

For additional information on Nero, see Richard Holland's *Nero, the Man behind the Myth* (Sutton Publishing, Stroud, 2000). Holland devotes an entire chapter to Agrippina and titles it "The Dominant Mother" (p 43 and further).

3. Building of the Coliseum (also spelled Colosseum and sometimes referred to as the Flavian Amphitheatre) only commenced in AD 70, shortly after Nero's death, and it was ultimately completed in AD 96. Given that modern minds associate gladiatorial pursuits with this iconic building, some poetic licence was taken with the reference.

PART 1 BEAUTIFUL LIES

Chapter 2: Science's most beautiful lie – Truth speaks for itself

4. See Gary Snyder "To the Children" at http://www.gratefulness.org/poetry/for_the_children.htm, accessed 28 July 2015.

5. The line emanates from Alfred Lord Tennyson's In Memoriam A. H. H., 1850, Canto 56, where the love of the Christian faith is juxtaposed with the wildness of nature:

   Who trusted God was love indeed
   And love Creation's final law
   Tho' Nature, red in tooth and claw
   With ravine, shriek'd against his creed

   Darwin's The Origin of the Species (1859) would soon be published, creating further focus on the dilemma. More recently, Richard Dawkins in his The Selfish Gene (1976) has been fond of the term: "It is ironic that Ashley Montagu should criticize Lorenz as a "direct descendant of the 'nature red in tooth and claw' thinkers of the nineteenth century .... As I understand Lorenz's view of evolution, he would be very much at one with Montagu in rejecting the implications of Tennyson's famous phrase. Unlike both of them, I think "nature red in tooth and claw" sums up our modern understanding of natural selection admirably". See https://goo.gl/NcR7wF, accessed 11 January 2016

6. See https://www.youtube.com/watch?v=RwUGSYDKUxU, accessed 28/12/2015

7. See Stephen Jay Gould "... we cannot understand history if we saddle the past with pejorative categories based on our bad habits for dividing continua into compartments of increasing worth towards the present. These errors apply to the vast paleontological history of life, as much as to the temporally trivial chronicle of human beings. I cringe every time I read that this failed business, or that defeated team, has become a dinosaur and is succumbing to progress. Dinosaur should be a term of praise, not opprobrium. Dinosaurs reigned for more than 100 million

years and died through no fault of their own; Homo sapiens is nowhere near a million years old, and has limited prospects, entirely self-imposed, for extended geological longevity." http://www.goodreads.com/author/quotes/19109.Stephen_Jay_Gould.

In his Nobel Peace Prize acceptance speech in 1964, Martin Luther King said: "I refuse to accept the view that mankind is so tragically bound to the starless midnight of racism and war that the bright daybreak of peace and brotherhood can never become a reality.... I believe that unarmed truth and unconditional love will have the final word."

8. See Plato's *Apology* (Benjamin Jowett's translation) at http://classics.mit.edu/Plato/apology.html. For further background on Socrates in the chapters that follow, see the *Crito* at http://classics.mit.edu/Plato/crito.html and the *Phaedo* at http://classics.mit.edu/Plato/phaedo.html, both translated by Jowett. While some effort was made to relate the thoughts of Socrates accurately, considerable licence was taken in using a far more modern idiom. All three websites accessed 28 July 2015.

9. The Gay Science, section 125 p 119 – 120 Cambridge University Press, edited by Bernard Williams, translated by Josefine Nauckhoff and Adrian del Caro – available online at http://www.holybooks.com/wp-content/uploads/The-Gay-Science-by-Friedrich-Nietzsche.pdf

10. For a discussion of the role of science in modern day jurisprudence in the USA, see Gatekeeping Soothsayers, Quacks, and Magicians: Defining Science in the Courtroom – Judging Science – Scientific Knowledge and the Federal Courts by Eileen Gay Jones in the 1999 William Mitchell Law Review, article 22, , Volume 25, Issue 1 at http://open.mitchellhamline.edu/cgi/viewcontent.cgi?article=1900&context=wmlr, accessed 28 December 2015

11. See Kenneth E. Boulding quoted by Richard Heinberg, in *The Party's Over: Oil, War and the Fate of Industrial Society* (New Society Publishers, Vancouver, 2003), p 43.

12. See http://www.oregonlive.com/books/index.ssf/2010/06/poetry_the_enlightenment_and_a.html, accessed 28 July 2015.

13. For a compelling discussion of rational vs romantic environmentalism, see Johann Hari in Slate.com "Move Over, Thoreau: Rationalist Environmentalism Better Prevail, and Fast", January 12 2009, available at http://www.slate.com/articles/arts/books/2009/01/move_over_thoreau.html, accessed 28 July 2015.

14. See http://www.bartleby.com/145/ww287.html, accessed 28 July 2015.

15. http://www.poets.org/viewmedia.php/prmMID/16083, accessed 28 July 2015.

16. Carl Edvard Marius Levy, obstetrician, professor and head of the Danish Maternity Institution in Copenhagen, see http://setsciencefree.org/historical-quotes-from-dogmatic-thinking/, accessed 28 July 2015.

17. http://www.nytimes.com/2003/10/07/health/the-doctor-who-made-his-students-wash-up.html?pagewanted=all&src=pm, accessed 21 June 2012. There are conflicting theories on the cause of his death. In practice this makes little difference: It is very clear that his theory was rejected and that he died in an asylum.

Chapter 4: The endgame of beautiful lies: warnings as fiction

18. Adapted from the well-known account in Aesop's Fables, see http://www.storyarts.org/library/aesops/stories/boy.html, accessed 28 July 2015.

19. The show can be heard online at http://archive.org/details/OrsonWellesMrBruns, accessed 28 July 2015.

20. The original script is available at http://www.sacred-texts.com/ufo/mars/wow.htm, accessed 28 December 2015

21. See William H. Davidow, "Overconnected: What the Digital Economy Says about Us", *Headline Business Plus*, 2011, pp 80–90.

22. Paramount Pictures, 1998, directed by Peter Weir; written by Andrew Niccol; co-starring Laura Linney, Noah Emmerich and Natasha McElhone.

23. Warner Brothers, 1999, directed by Larry and Andy Wachowski; starring Keanu Reeves, Laurence Fishburne and Carrie-Anne Moss.

24. Warner Brothers, 2010, written and directed by Christopher Nolan, starring Leonardo Di Caprio, Marion Cotillard and Michael Caine.

PART 2 UGLY TRUTHS

Chapter 6: A new serial killer on the loose/A scientific report on PECCA's

25. Tom Kington, "Pavarotti Mimed at Final Performance", *The Guardian* (London), 7 April 2008, at http://music.guardian.co.uk/classical/story/0,,2271470,00.html, accessed 7 July 2012.

26. See Frank Rich, "Two Top Guns Shoot Blanks", New York Times, 19 June 2005, available at http://www.nytimes.com/2005/06/19/opinion/19rich.html?pagewanted=print&_r=0, accessed 6 March 2014

27. An issue is policy related when it requires political support for its resolution. This again implies that, in democracies, the majority of voters in the perception of their leaders will think it's a good idea to do what is required to resolve it. In the hardest cases, international political support can be required from different countries with different political systems and different levels of affluence.

28. The term "environomic" is used to denote something having inter-linked components of both environment and economy. If a country has a balance of payment deficit, that is an economic problem without an obvious environmental aspect to it. Conversely, if people leave behind plastic bottles and papers in a sports stadium after a game, that is (within that narrow context) an environmental problem without a significant economic aspect to it. In the combined sense, an oil spill in the ocean where there is a good fish resource is environomic – the

environment is harmed and the people who depend on catching fish suffer economic harm. The costs/benefits encompass both the environmental and economic realms. An issue cannot be environomic within the context of this book if an economic cost-benefit analysis of its resolution is overwhelmingly positive when measured in monetary terms. To use an extreme example, the environmental pollution of an empty parking lot "polluted" with gold coins would not be environomic because people would rush to pick up the coins and make a profit.

To use another example, if (in a dry area) water is being lost because someone forgot to close the sluices of a dam, the cost of pushing the button and closing the sluices is negligible, while the benefit of no longer losing the water is significant. The most intractable examples of environomic disputes often involve cases where money has to be spent in order to safeguard or promote things that are not easily expressed in monetary terms. So for instance, should the developer of a housing complex be required to spend an extra million dollars to prevent the extinction of a rare species of owl, for which that piece of land constitutes its last surviving home? Also in the category of a difficult environomic matters would be the case where significant expense might have to be incurred to prevent damage much later; and moreover, there not being complete certainty about whether the damage is likely to occur – and if it occurs, how serious it is likely to be? As an example, a man living near a river is told that he has to spend the equivalent of 20% of the value of his property every year to prevent soil erosion on the banks of the river. This is because the banks of the river will erode and the river will broaden dramatically when there are heavy thunder storms. He is told that if he does not follow this procedure the foundations of his house will crack within three years and his house will start tumbling into the river within a decade. But one of his friends who has studied geology insists that this scenario will proceed much slower than predicted, and that the river will never come closer than twenty yards of his home.

29. In the environomic context, a problem can be complex because the science is intricate and/or because the resolution requires drastic changes to the societal status quo. Examples of non-qualifying issues would be having a smoke emitting car with a defective exhaust (it is clear to everyone it is polluting) or an effort to diminish the throwing of

litter next to a highway (it doesn't require a big change to the societal status quo to stop littering). An example of a complex issue would be outlawing smoking in public places (the science of proving the dangers of secondary smoke inhalation is intricate) and moreover it is generally thought by smokers to have a considerable impact on their lives (drastic change to the societal status quo)

30. If a problem requires collective action for its peaceful resolution that means that no single party, however well-meaning, can peacefully resolve it on her/his own. In extreme cases the collaboration of many international entities with very different agendas can be required. The international climate negotiations have seen close to 200 countries negotiate for more than twenty years.

31. When we burn fossil fuels (burn coal to generate electricity, drive a car, cook on the gas stove) the result is that carbon dioxide is emitted into the earth's atmosphere. Carbon dioxide is a very inert gas, which means it stays in the atmosphere for decades. There are other gases showing similar attributes. These include methane and nitrous oxide that are emitted by other human activities like agriculture, the building of very large dams and the making of fertilizer.

    These gases have been building up in the atmosphere from a level of about 280 parts per million before the Industrial Revolution (1750 – 1850) to about 394.45 PPM in March 2012. At the same time, changes have been observed in the global environment: Glaciers have been melting/retreating; the Arctic sea ice has shrunk; rainfall patterns have been changing; the global temperature has on average been increasing; there has been an increase in extreme weather events like cyclones; the seasons have changed and plants and species are migrating to higher areas or areas further from the equator.

    The most common name for the phenomenon is "global warming", but the impacts are felt far wider than just increased average temperatures. Tropical diseases occur in new places. Crops fail. In time, areas will become uninhabitable and there will be environmental refugees.

    The reason is simply that too much greenhouse gas is being emitted – more than the earth can naturally deal with. The more parts per million we allow to build up in the atmosphere, the warmer the planet will become, the less stable the climate will be, the more extreme and less predictable the climate will be. Sea levels will rise and threaten low-lying areas.

In order to address this problem we would have to rapidly and fundamentally change the way we live, by (amongst other things) using energy sources other than the cheap fossil fuels we have been used to since the Industrial Revolution. We would certainly have to switch to renewable energy and possibly also to nuclear power. We would have to become dramatically more efficient in our energy use (eg smaller cars). More generally we need to consume less, travel less, and produce more locally. It would help to eat less meat and accept that our economic growth ultimately will be higher if rises more gradually (rather than very quickly and then reverse itself due to run-away climate change).

32. If we start to investigate the PECCA status of climate change, we see that it is clearly Policy related (P): The things that cause climate change require governmental and international action to change. If I switch to a small car but companies keep giving subsidies to their employees to drive big cars because of the tax regime, few people will join me. The policy framework must encourage or mandate a change. Climate change is clearly Environomic (E) because it has both environmental and economic components and it has not been clear that it is less expensive to avoid climate change than to allow it to develop. Indeed, the costs are incurred now, the benefits occur far into the future, because of the time value of money concept, long debates can be had about where the cost benefit actually lies. If we assume that money now is worth much more than money later, we will more easily call a halt to spending on avoiding climate change. The converse is also true. Climate change is Complex (C) for two reasons: the science is complex and the adjustments to human behaviour required to address biodiversity are profound – the lifestyle changes mentioned above would be difficult to make for many people. It is clear from the above that Collective Action (CA) is required to address climate change peacefully. Notionally, a few large polluters can frustrate the efforts of many conscientious people trying to do the right thing.

33. On a planetary level, biodiversity can be defined simply as the degree of variation of life forms, measured amongst other ways in the quantum of species and the amount of variations within species . It is largely a numbers game: how many species do we have?

Within each species, how many variations are there? Regarding the number of species in a given area, it is a case of "the more the merrier": the more species there are, the better they all tend to do. For this reason, biodiversity is a way to measure the health of the planetary ecosystem. Many species equal good health.

According to authoritative sources, the total number of species could be in the tens of millions, with insects constituting by far the great majority, followed by bacteria and fungi (both probably in the millions of species). Most insect species have probably not been discovered as yet.

Biodiversity is important to us, because on a more abstract level, it measures how secure continued human survival on our planet is: If we accept that a greater number of species generally means better health for all, then good biodiversity is like a substantial pension plan: We can be confident that we will live well into the future. Expressed more tangibly, biodiversity is of monetary and convenience value to humans.

To use a simple example, pig farmers often use anaerobic lagoons to treat pig slurry. They pump the pig manure suspended in the slurry into a lagoon that is starved of oxygen and is thus "anaerobic". Over the course of about 3 – 4 weeks, anaerobic organisms break down the pig manure into more useful and less harmful substances – free of charge. Methane rich gas is one by-product that can be used for energy generation or heating.

In a similar way biodiversity supports air quality, water purification through natural wetlands, and pollination through bees and other organisms – to mention only a few examples. The economic value of 17 ecosystem services for Earth's biosphere has an estimated value of more than USD 30 trillion per year. This exceeds Microsoft's annual operating income by a factor of roughly 1,000.

Conversely, poor biodiversity equates to human risk. So, for instance, in the 1840s in Ireland, potato was a staple food. Only two varieties were planted. When the potatoes were attacked by a disease and the potato crop failed, a million people died of starvation and another million emigrated. Poor biodiversity was thus a direct contributor to the Irish diaspora. Had more species been planted it is likely some would have been resistant to the disease and the effects would have been less severe.

Today just three crops – maize, rice and wheat – supply nearly half of the world's food. Within these foods, diversity is rapidly declining. So for instance, China had 46,000 rice varieties in the 1950s, but this has dropped to about 1,000.

It is clear that biodiversity is of great importance to us.

The many threats to biodiversity include pressures from habitat loss, land use change and degradation, unsustainable water use and the spread in invasive alien species in areas where they did not exist before. Most of the above is heavily coincident with human activity: It is people that sometimes slash and burn down an old-growth forest to make space for agriculture and in the process diminish the space within animals and plants can live. It is people that use water unsustainably and it is us that spread plants and animals to new parts of the globe – advertently and inadvertently, legally and illegally. Biodiversity loss is thus within our hands to prevent.

See Costanza, R.; d'Arge, R.; de Groot, R.; Farberk, S.; Grasso, M.; Hannon, B.; et al., Karin; Naeem, Shahid et al (1997). "The value of the world's ecosystem services and natural capital". Nature 387 (6630): 253–260. Bibcode 1997Natur.387..253C. doi:10.1038/387253a0. http://www.uvm.edu/giee/publications/Nature_Paper.pdf. See also p 11 of the Global Biodiversity Outlook 3 at http://www.cbd.int/doc/publications/gbo/gbo3-final-en.pdf (accessed 5 May 2012) "Better protection of biodiversity should be seen as a prudent and cost-effective investment in risk-avoidance for the global community. The consequences of abrupt ecosystem changes on a large scale affect human security to such an extent, that it is rational to minimize the risk of triggering them – even if we are not clear about the precise probability that they will occur. Ecosystem degradation, and the consequent loss of ecosystem services, has been identified as one of the main sources of disaster risk. Investment in resilient and diverse ecosystems, able to withstand the multiple pressures they are subjected to, may be the best-value insurance policy yet devised".

34. If we start to investigate the PECCA status of biodiversity, we see that it is clearly Policy related (P): The things that destroy biodiversity require governmental and international action to change. Each farmer wanting to open up a forest for agriculture may be acting in good faith and may have only a minor impact, but the collective impact is considerable and can only be avoided through appropriate policy that is properly policed. Biodiversity is clearly Environomic (E) because it has both environmental and economic components and it has not been clear that it is less expensive to maintain biodiversity than to deplete it: Indeed, the farmer is almost certain to make more from his planted crops than he would from a pristine forest. Biodiversity is Complex (C) for two reasons: the science is complex and the adjustments to human behaviour required to address biodiversity are profound (how for instance do you explain the logic of not planting to the farmer with the forest). It is clear from the above that Collective Action (CA) is required to address biodiversity

35. The following international, multi-lateral agreements have been entered into regarding climate change:

    - The United Nations Framework Convention on Climate Change entered force on 21 March 1994 and has been agreed to by 195 countries – practically the entire global community. The aim of the convention is to stabilise greenhouse gas concentrations in the atmosphere at a "level that would prevent dangerous anthropogenic (man-made) interference with the climate system". Several international bodies have been created by the convention, including the Conference of the Parties (COP). This is a body consisting of all governments that have ratified the treaty that governs everything under the treaty. The COP meets about once a year and has done so 21 times so far. Also created under the auspices of the UNFCCC is the Intergovernmental Panel on Climate Change or IPCC – a scientific body for the assessment of climate change consisting of thousands of scientists and set up under the leadership of the United Nations Environment Programme (UNEP) and the World Meteorological Organisation (WMO) and blessed by the UN General Assembly. The UNFCCC also has a secretariat situated in Bonn, Germany, employing about 300 people fulltime in the administration of the convention. The scale of activity is well reflected by the extensive UNFCCC website.

- The Kyoto Protocol is an international agreement that emanated from the UNFCCC. It contains binding targets to some countries for diminishing their emissions of greenhouse gases in the period 2008 – 2012. It has been agreed to by 191 countries and entered force in February 2005.
- Under the Kyoto Protocol, the so-called flexible mechanisms have been created. These include Joint Implementation and the Clean Development Mechanism or CDM. These mechanisms are aimed at fostering international co-operation and technology transfer to ensure that emissions reductions take place in the most cost-effective manner. The CDM on its own has led to more than 4,000 strictly audited, international projects in renewable energy, methane capture, energy efficiency and other technological interventions that collectively will lead to the avoidance of more than 2 billion tonnes of carbon dioxide equivalent. This is approximately equal to South Africa's total emissions for five calendar years. All these projects are certified by the host countries to be assisting them to achieve sustainable development.
- An international, voluntary, public-private partnership between Australia, Canada, India, Japan, the People's Republic of China, South Korea, and the United States called the Asia-Pacific Partnership on Clean Development and Climate was announced in July, 2005 and launched in January 2006. It is now defunct.
- Many other organisations have sprung up and conferences been organised. Amongst these are Business Action on Climate Change, the World People's Conference on Climate Change and the 2011 United Nations Climate Change Conference.
- The Climate Action Network or CAN is "a worldwide network of over 950 Non-Governmental Organisations (NGOs) in over 90 countries working to promote government and individual action to limit human-induced climate change to ecologically sustainable levels".

In summary it is clear that the international effort on climate change has been Herculean.

If success is measured by preventing the build-up of greenhouse gases in the atmosphere, the huge and very expensive effort has yielded very little. The July 2015 greenhouse gas concentration in the atmosphere was 402.8 PPM. In 2011 it was 392.4 and another year before that, 391.08. See http://co2now.org/, accessed 6 May 2012. Professor Clive Hamilton investigated the matter with reference to extensive scientific research and concluded that we may already be irrevocably on our way to 650 PPM. See C. Hamilton, Requiem for a Species: Why We Resist the Truth about Climate Change (Earthscan, London, 2010), pp 15–21. In the eight years after the Kyoto Protocol was agreed, energy related emissions had grown by 24%. From the year 2000, carbon emissions have grown at about 3% per annum, which is worse than the IPCC's worst case scenario modelled a few years before. There is a note of anger and desperation in Hamilton's writing that is not uncommon in books dealing with climate change.

On the United Nations Website, the point is made that climate change is starting to impact other development initiatives like the Millennium Development Goals:

"Climate change threatens to erode the gains made through humanitarian and development efforts to improve the lives and strengthen the self-sufficiency of billions of people, and to reduce their vulnerability to a myriad of natural, economic and social threats. The Millennium Development Goals represent the world's collective effort to strive for a fairer world. Looking at the current and projected impacts on climate change– for example, on agriculture, freshwater and human health– one could easily identify where climate change impacts might challenge the progress of the MDGs".

To quote from the United Nations website for the international convention dealing with climate change:

- The average temperature of the earth's surface has risen by 0.74°C since the late 1800s. It is expected to go up another 1.8°C to 4°C by the year 2100 if no action is taken. That's a fast and intense change in geological time. Even if it "only" gets another 1.8°C hotter, it would be a larger increase in temperature than any century-long trend in the last 10,000 years.

- About 20-30% of plant and animal species is likely at higher risk of extinction if the global average temperature goes up by more than 1.5 to 2.5°C.
- Nine of the last 10 years were the hottest years on record, according to the United States' National Oceanic and Atmospheric Administration (NOAA). 2005 and 2010 tied for first place. In second place was 1998.
- The average sea level rose by 10 to 20 cm over the 20th century. An additional increase of 18 to 59 cm is expected by the year 2100. Higher temperatures cause ocean volume to expand. Melting glaciers and ice caps add more water. And as the bright white of ice and snow give way to dark sea green, less and less rays from the sun are reflected back into space, intensifying the heating.

If one studies the document and the Stern Review, it is clear from the content that climate change is important because, as a global population, if allowed to continue:

- It will probably cause us to have less food;
- It will almost certainly cause us to have less water;
- It will affect us through diseases that previously only occurred in other areas (malaria and other vector-borne diseases in particular);
- we will have to bear an increase in weather-related disasters that will generally be more extreme;
- we will see some of the places where we currently live become inhabitable;
- a great many plants and animal species will become extinct.
- It will cost us between 5 and 20% of GDP.

In 2009, UN Secetrary General Ban Ki-moon said the following during his keynote speech to the World Federation of U.N. Associations:

> "We have less than 10 years to halt the global rise in greenhouse gas emissions if we are to avoid catastrophic consequences for people and the planet. It is, simply, the greatest collective challenge we face as a human family."

To the UN Security Council, he said:

> "We must make no mistake. The facts are clear: climate change is real; it is accelerating in a dangerous manner; and it not only exacerbates threats to international peace and security, it is a threat to international peace and security. Extreme weather events continue to grow more frequent and intense in rich and poor countries alike, not only devastating lives, but also infrastructure, institutions, and budgets – an unholy brew which can create dangerous security vacuums."

The Pentagon seems to agree. In 2010 it was reported the Pentagon would from that point on rank global warming as a destabilizing force, adding fuel to conflict and putting US troops at risk around the world. At the time it was said that more than 30 US bases are threatened by rising sea levels.

In conclusion then: climate change, despite intense efforts to address it, is not under control.

36. It is clear that a tremendous amount has been done internationally to conserve biodiversity. Just about all countries have acted jointly for two decades or more to address this PECCA. Moreover, the United Nations has decided that 2011-2020 would be the UN Decade on Biodiversity. The issue could hardly be receiving more attention.

The following international, multi-lateral agreements have been entered into regarding biodiversity:

United Nations Convention on Biological Diversity (1992). This agreement entered force on 29 December 1993. It aims to conserve biological diversity, foster sustainable use of its components and ensure fair and equitable sharing of benefits arising from genetic resources. It has presently been signed by 193 countries. Several international bodies have been created by the convention, including the Conference of the Parties (COP). This is a body consisting of all governments (and regional economic integration organisations) that have ratified the treaty that governs everything under the treaty. The COP meets about once a year and (at the time of the first draft of writing) has done so 10 times so far. It can create expert committees, periodically review progress and collaborate with other international organisations and agreements. Thus, about once a year, thousands of government and NGO people travel to a single destination to discuss progress. In between a great many work on nothing else. The level of activity is clearly perceived by looking at the website.

The Cartagena Protocol on Biosafety is an "addendum" to the Biodiversity Convention mentioned above and aims primarily to prevent adverse effects that could result from living modified organisms (LMO's) emanating from modern biotechnology. It entered into force on 11 September 2003.

The Convention on International Trade in Endangered Species (CITES) was mentioned in Chapter 4, in the discussion of elephants in Africa. To recap, the aim is to prevent the international trade in specimens of wild animals and plants from compromising the survival of the species in the wild. There are more than 25,000 species of plants and animals protected by CITES. The agreement has been signed by 175 countries.

The Convention on Wetlands of International Importance (the Ramsar Convention) is an international agreement for the conservation and sustainable utilisation of wetlands. It has been agreed by 160 countries and been in force since December 1975. Like the Biodiversity Convention, there is a periodic Conference of the Parties (a global meeting) with all the attendant travel, expense and effort that entails. There have been 10 meetings of the COP at the time of writing.

The Bonn Convention on Migratory Species (Bonn Convention) aims to conserve terrestrial, marine and avian migratory species throughout their range. It has been agreed by 116 countries and been in force since 1983. The considerable level of activity is evident from a look at the website.

The state of play on biodiversity is very clearly demonstrated by the Global Biodiversity Outlook, which is published periodically by the Secretariat of the Convention on Biological Diversity. The third version appeared in 2010.

In the foreword, the Secretary General of the UN, BAN Ki-moon, says the following:

> "In 2002, the world's leaders agreed to achieve a significant reduction in the rate of biodiversity loss by 2010. Having reviewed all available evidence, including national reports submitted by Parties, this third edition of the Global Biodiversity Outlook concludes that the target has not been met. Moreover, the Outlook warns, the principal pressures leading to biodiversity loss are not just constant but are, in some cases, intensifying...The consequences of this collective failure, if it is not quickly corrected, will be severe for us all. Biodiversity underpins the functioning of the ecosystems on which we depend for food and fresh water, health and recreation, and protection from natural disasters...Its loss also affects us culturally and spiritually...This may be more difficult to quantify, but is nonetheless integral to our well-being. Current trends are bringing us closer to a number of potential tipping points that would catastrophically reduce the capacity of ecosystems to provide these essential services. The poor, who tend to be most immediately dependent on them, would suffer first and most severely. At stake are the principal objectives outlined in the Millennium Development Goals: food security, poverty eradication and a healthier population."

Just a few facts from the executive summary show why there should be so much concern:

- Species which have been assessed for extinction risk are on average moving closer to extinction.
- Nearly a quarter of plant species are estimated to be threatened with extinction.
- The abundance of vertebrate species, based on assessed populations, fell by nearly a third on average between 1970 and 2006, and continue to fall globally.
- Crop and livestock genetic diversity continues to decline in agricultural systems.

In conclusion, despite intense efforts to address it, biodiversity loss is not under control

37. Present Progress in Resolving Different Kinds of Environmental Challenges

Case study	Acronym	Policy related ("P")	Environomic ("E")	Complex ("C")	Collective action ("CA")	Present outcome
Organic food	ECCA	×	√	√	√	√
Ozone depletion	PCCA	√	×	√	√	√
Elephant extinction in Botswana	PECA	√	√	×	√	√
Safe toys	PEC	√	√	√	×	√
Biodiversity	PECCA	√	√	√	√	×
Climate change	PECCA	√	√	√	√	×

38. ECCAs: Leveraging the benefits of organic farming

In the search for success stories containing three of the four key criteria, the first category comprises the Environomic (E) problem that is Complex (C) and requires Collective Action (CA), but has no Policy (P) component.

One's thoughts immediately go to Corporate Social Responsibility (CSR) and the question of what successes can be shown that were achieved in the absence of any government policy or legislation enforcing the change. There are many CSR success stories, but very few (if any) that comprise an entire industry. A few leading players may have blazed trails and set standards about sustainability, but when this was done in the absence of industry-wide action, the question always arises whether better branding and product differentiation was a strong driver, leading to better profits, which would in turn have meant that we are not really talking about an environomic problem.

The best example in the category turns out to be the consumer-driven move from industrial agriculture to organic food. Industrial agriculture has been evolving ever since the Industrial Revolution. From this time onwards, synthetic fertilizers such as phosphorus, potassium and nitrogen began to be used to enhance plant growth. Yields improved and agriculture has become more intensive. Various discoveries have allowed livestock to be raised indoors. Synthetic pesticides were also developed. In recent times, genetically modified organisms have been developed to resist pests and droughts. Farms have become larger and owned by fewer, larger entities that are able to apply economies of scale. All the above have led to dramatic increases in food production that in turn are consumed by the fast-growing global population.

However, there is a significant downside to industrial agriculture. Amongst the adverse effects are very intensive water usage; highly intensive fossil fuel usage; potential groundwater pollution by fertilizers/ insecticides; resistance developing in the target organisms of insecticides; job losses; deteriorating rural living conditions; and the possible spread of new diseases between humans and animals (Mad Cow Disease for example).

The counterpoint to industrial agriculture has been the return over the past sixty years to organic farming and organic food. This means that the food is produced using methods that do not involve (amongst other things) any of the practices mentioned above, like chemical fertilizers or pesticides, nor the planting of genetically modified organisms. Energy and water efficiency is higher than would be the case for industrial agriculture.

Another positive effect is the superior conservation of diverse ecosystems.

The organic production of food is *environomic* because, as will be seen below, the cost of food produced organically is usually higher and the purely economic cost-benefit analyses of organic farming (ignoring externalities) over industrial agriculture would thus be negative. It is also naturally *complex*, with many invisible and subtle changes to the environment being the result of industrial agriculture. From the consumer side, the issue also demands *collective action* as a market of a size greater than the critical threshold has to be established and then grown in order to sustain the growth of organic farming. The industry is driven by consumers who generally pay more for their food because it is organic.

Organic products typically cost 10% to 40% more than similar conventionally produced products. On a global basis, the organic food market has nevertheless been growing at about 20% per annum for more than a decade and is presently worth more than US$50 billion per annum.

Regulations rigidly control what can be defined and sold as organic food but there is no discernible global policy initiative to promote it over food produced by industrial agriculture. The success thus seems to be consumer driven and is an example of how an *environomic* issue that is complex and that requires *collective action* can nevertheless be improved even in the absence of *policy* support.

There are very few examples of ECCAs that have been resolved – it appears generally that if the benefit-cost is negative and there are no laws about the issue, the cheap and environmentally harmful option tends to be chosen. Some companies and individuals are responsible to the extent of being willing to pay to do the right thing, but entire industries seldom are. In organic food production there is at least one example of successful action on an ECCA.

39. PECs: Removing toxic children's toys from circulation

The final case study that deals with instances where progress is being made with challenges displaying three of the four critical criteria, is one that is Policy (P) related, Environomic (E) and Complex (C), but does not require Collective Action (CA). In this example, the reason why no *collective action* is required is because the harmful activity emanates from a single source.

Mattel Inc., based in California, is the world's largest toy company. Founded in 1945, it produces Barbie dolls and Matchbox toys, to name just two illustrious brands. Over time, production was outsourced entirely to China. As part of the production process, some toys are painted.

Increased levels of lead in the body causes lead poisoning. The result is extremely adverse impacts on health which may potentially affect the nervous system, the heart and the kidneys, to mention just a few.

Mattel's toys being manufactured in China and painted with a lead-based paint had the potential to cause a significant health hazard to children. In some cases the US Federal limit in parts per million was exceeded by more than 10,000%. The result was that in 2007, Mattel recalled more than a million toys internationally. While lawsuits and fines resulted, the problem had clearly been addressed even before that.

The danger of lead poisoning through the paint on children's toys is in need of regulation – thus it has a *policy* component to it. It is also *environomic* and *complex*. Once these three elements are in place (as they were in the Mattel case), the problem can however be resolved without *collective action*. Mattel recalled the toys and the resolution of the problem was set in motion. We can deal with PECs.

40. PCCAs: Reversing ozone depletion

Moving on in the discussion of challenges displaying three of the four critical criteria, we discuss PCCAs. These are challenges having a Policy (P) component; that are Complex (C); and that require Collective Action (CA) for their peaceful resolution – but are not Environomic (E). This discussion leads us to protection of the ozone layer.

In the earth's atmosphere there is a layer of ozone which protects us against the ultraviolet-B radiation reaching the earth from the sun. If this were absent or depleted, a variety of adverse effects could result, ranging from skin cancer to crop damage, to negative impacts on marine phytoplankton.

In the early 1970s, scientists hypothesized that chlorofluorocarbon ("CFC") chemicals generally used in refrigeration and aerosol cans were harming the ozone layer. To make matters worse, the offending particles – once emitted – would continue to harm the ozone layer for 50 years or more thereafter.

The ozone story is deeply policy-related. Rowland and Molina testified on the matter before the United States House of Representatives in December 1974. More funding was made available to study the issue. In 1976, the US National Academy of Sciences threw its weight behind the scientific credibility of the ozone depletion hypothesis.

Because the science behind ozone depletion is so complex, initially divergent opinions were easy to hold. The CFC industry was largely led by the chemical giant DuPont. In the early days, the chair of the board of DuPont was quoted as saying that ozone depletion theory is "a science fiction tale ... a load of rubbish ... utter nonsense". The chancellor of the university where Rowland worked received a written complaint from the company promoting the first aerosol spray can valve.

For the next nine years nothing much happened. There were very credible scientific indications that a potentially destructive environmental practice was occurring on large scale, but no meaningful action was taken to curb it. Then, in 1985, the "hole in the ozone layer" was discovered. Three British scientists, Farman, Gardiner and Shanklin found abnormally low ozone concentrations near the South Pole in Antarctica, and immediately tied this empirical result to the CFC hypothesis. NASA used satellite recordings to confirm the finding. Suddenly the public was convinced and the politicians were galvanized. Very soon afterwards the Vienna Convention was signed. This was a framework agreement without binding, legally enforceable clauses, but such clauses were added within eighteen months thereafter by the Montreal Protocol. Both international agreements have been ratified by 196 countries plus the European Union – all the United Nations members.

In the time between the signing of the two agreements, some industry voices were fighting a rearguard action. In 1986, the Alliance for Responsible CFC Policy (an association representing the CFC industry founded by DuPont) was still arguing that the science was too uncertain to justify any action. In 1987, DuPont testified before the US

Congress that "we believe that there is no immediate crisis that demands unilateral regulation". But the pendulum had swung decisively towards acceptance of the link between CFCs and ozone depletion and there was no turning back.

In recognition of their work on a complex field of science, Paul J. Crutzen, Mario Molina and Frank Sherwood Rowland received a Nobel Prize for Chemistry.

It is thus clear that the issue of ozone depletion is *complex*. Reversing ozone depletion is by its nature also something requiring *collective action* for its peaceful resolution. If some countries continued using CFCs, the problem would persist even if other countries stopped doing so. The uniform adoption of the two international conventions dealing with the problem show that *collective action* was achieved.

An investigation of the action taken shows that preventing ozone depletion may be *policy related*, *complex* and in need of *collective action*, but it doesn't quite have the attribute of being an *environomic* issue. This is because the potential harm of ozone depletion completely dwarfs the relatively low cost of containing the problem.

The way in which the problem has been dealt with since the mid-eighties confirms this: A multi-lateral agency was set up and a timetable agreed for the phasing out the CFCs. The fund is capitalized tri-annually – pledges for the period 1991 to 2005 amounted to US$2.1 billion. At an average of about US$130 million per annum this amounted to about 0.00025% of global GDP. Calculated differently, the budget for that fifteen- year period equaled what just one industrialized country presently spends every five days to cover the medical costs of obesity.

As against this, ozone depletion in the same country could cause an additional 100,000 extra cases of skin cancer every year by 2050. The cost of non-melanoma skin cancer in that country already exceeds the entire annual, global cost of reversing ozone depletion by a factor of 3:1.

The Montreal Protocol has been very successful in addressing the ozone problem. Not only has it been universally endorsed and ratified by the international community, the deterioration in the ozone layer has seemingly been arrested. By 2009, the use of 98% of the harmful substances had been terminated. It is projected that by 2050, the pre-1980 levels of atmospheric ozone can be achieved again.

The World Meteorological Organization's 2006 scientific evaluation of the effects of the Montreal Protocol states: "The Montreal Protocol is working: There is clear evidence of a decrease in the atmospheric burden of ozone-depleting substances and some early signs of stratospheric ozone recovery". Kofi Annan, former Secretary General of the UN, has called it "perhaps the single most successful international agreement to date".

Economically, it may have been a no-brainer to address ozone depletion, and in this sense it may not have been a true *environomic* challenge. Resounding success has however been achieved with a highly complex challenge that had a large policy component and that required collective, international action.

We can cope with formidable PCCAs.

41. PECAs: Preventing elephant extinction in Botswana

When we discuss species extinction, elephants are likely to enter the conversation earlier rather than later. With the *Mammuthus* genus already extinct, the African and Indian elephants form part of the group of "charismatic" animals that people would like to save: They are large, mammalian and credited with exceptional intelligence. In their natural environment they have no natural enemy and are largely peaceful herbivores. They form deep, long-term family bonds which between a mother and daughter can be lifelong.

The ivory in their tusks has been coveted internationally for centuries, leading to extinction of elephants in North Africa as far back as a millennium ago. The total number of African elephants dropped from 1.3 million in 1979 to about 600,000 in 1989 to about 470,000 in 2010.

Elephant protection requires Policy (P) support – regulations that outlaw excessive elephant hunting and that are enforced. The issue is by its very nature Environomic (E) – the aim is to prevent the extinction of a beloved animal species that plays an important role in the biosphere. The economic value of the ivory is the counterpoint to this and creates the inherent tension between elephant conservation and elephant resource utilization. In the third place, elephant conservation requires Collective Action (CA) – if the people living near the conservation areas don't support the initiative but rather join the poachers, the initiative will fail.

Elephant conservation is however not Complex (C) as defined above – the science for a start is (in principle) simple: shoot too many elephants and they become extinct. Moreover, conserving African elephants does not in principle require significant societal change – a small minority hunt/poach them and the main market for ivory is in Asia and Europe.

The Convention on International Trade in Endangered Species of Wild Fauna and Flora (CITES) deals elaborately with the ivory trade and has for periods enforced a total ban on ivory trade in an effort to protect elephants. This is the *policy* support component. Botswana is a signatory and despite its stated opposition to a total ban on ivory trade has been diligent in enforcing the ban and managing the elephant population in Botswana's game reserves, mostly Chobe. Hunting was confined to very special circumstances and anti-poaching units were established around the country. In this manner *collective action* was ensured.

The result has been dramatic – in 1960 the Botswana elephant population was estimated at 8,000. By 2003 this had grown to 123,000. This success has come as a mixed blessing, as we will see below, but it does prove that it is possible to resolve PECAs: *policy* related, *environomic* problems that require *collective action* but that are not *complex*.

By way of caution, however, it needs to be added that the investigation above has not taken into account the balance in the ecosystem within which these elephants live. It is arguable that this component adds complexity and elevates the matter to PECCA status and a high degree of intractability.

Chapter 8: What the people on the yacht want to hear

42. See "Pride, Prejudice and Happiness: Readers Choose Favourite Endings", by John Ezard, The Guardian, 2 March 2006, at http://www.theguardian.com/uk/2006/mar/02/books.booksnews, accessed 7 March 2014.

43. "Box-office statistics prove that no matter how inartistic a happy ending picture may be, it will always get more box office than a more artistic unhappy ending picture" See Herman Lewis as quoted by Eran Preiss, in, "Not such a Happy Ending: The Ideology of the Open Ending", Journal of Film and Video, Vol. 42, No. 3, Fall 1990

44. Available at www.disneydvds.co.uk and wholly recommended.

45. Published by Secker & Warburg, London, 1994

46. By Studio Canal and Working Title Films, 2001, directed by John Madden; produced by Tim Bevan, Eric Fellner, Mark Huffham and Kevin Loader. It starred Nicolas Cage as Corelli and Penélope Cruz as his Pelagia.

47. See "The Writer's Atonement" by Melinda Robb, where the possible errors that may arise from the dual filters of writer's perception and reader's perception are eloquently discussed. Available online at ojs.library.dal.ca/index.php/verso/article/download/537/553

Chapter 10: Green Sysiphus

48. In Greek mythology, Sisyphus was an iron-fisted, yet shrewd character, who refused to lie down and accept his place in the pecking order. For this he would eventually be punished very heavily.

    Let's recount his refusal to yield.

    He was a king from a prominent family. He had a progressive side to him, promoting navigation and commerce – but he also had a dark side. He ruled mercilessly over his subjects, had a weakness for money and was dishonest.

Up to this point, alas, he'd done nothing that was sufficiently remarkable even to achieve infamy. But then one has to choose one's enemies with care.

In the ancient Greek world, hospitality to guests and travellers was very important – so important it was watched over by Zeus himself. There were certain rituals that needed to be observed as part of the "Xenia" or the rules of hospitality. The host had to respect the guest, offering him/her food, drink and a bath without interrogating him/her personally. Likewise, the guest was required to respect the host. And the host was supposed to present a parting gift to the guest.

The reason why Xenia was so important was that you never knew which god might be coming to your door in disguise.

The importance of hospitality can be seen in the fact that its non-observance was responsible for more than one war. So for example the Trojan War started began when a host abducted his guest's wife (Helen of Troy).

It all started to go wrong for Sisyphus when he failed to observe the Xenia and killed travellers and guests. As the guardian of Xenia, this upset Zeus deeply. Sisyphus then started showing the conniving side of his character, even consulting the Oracle of Delphi to understand how he could kill his foe Salmoneus without any consequences to himself. He seduced his niece in a plan to spawn offspring that could dethrone Salmoneus. Later, Sisyphus told one of Zeus's closely guarded secrets to the river god Asopus, in exchange for a favour.

Zeus ran out of patience and ordered the god of death to chain King Sisyphus in the underworld. But Sysiphus was irrepressible – in a preview of the latter-day robber handcuffing the cop with his own handcuffs, Sisyphus chained Thanatos to the underworld and escaped, thus incapacitating the very machinery of death not just for himself but for all humanity.

Faced with this outrage the gods stood together, freed Thanatos, captured Sisyphus and eventually restored the normal order of things. Sisyphus was ordered to be killed, but still he refused to yield. Just before dying he convinced his wife not to bury him with respect but to throw his body in the town square. Using this humiliation that begged for correction as an excuse, Sisyphus then pleaded with Zeus's daughter Persephone, queen of the underworld, to let him return to the upper world. To life. He convinced her and escaped again.

Having upset just about everyone who mattered, Sisyphus came to the end of his resourcefulness. His luck ran out. A terrible punishment had to be reserved for him, for he'd shown a complete lack of respect.

Zeus finally nailed Sisyphus through the intervention of Queen Persephone, who condemned Sisyphus to the eternal task of rolling a huge boulder up a hill. And every time, just before the top, the boulder rolled down to the base of the hill, and Sisyphus had to walk down and start all over again.

49. Albert Camus was very interested in how man finds meaning in a world that he perceives as meaningless, absurd, and devoid of hope. In discussing this in *The Myth of Sisyphus*, first published in 1942, now reprinted by Penguin (London, 2005), Camus found in Sisyphus the perfect metaphor for modern man's existence. His daily task is devoid of meaning and hope. As he labours up the mountain, he knows that nothing will be achieved, but he does it nevertheless. In an approach that shows similarities with the approach of psychologist and holocaust survivor Victor Frankl's *Man's Search for Meaning: An Introduction to Logotherapy* (1962), Camus indicates that Sisyphus (and modern man) has control over how he reacts to his situation. He could reject his suffering and commit suicide. Camus, however, feels that the proper response is to revolt. For this reason, in *The Myth of Sisyphus*, he says that when Sisyphus walks down the mountain, we should imagine him to be happy.

    Camus's legacy has been a philosophy of dealing with an absurd world. *The Myth of Sisyphus* is a primary pillar in this body of work.

50. See http://www.the-testament-of-truth.com/noah/content/set3.htm and http://www.theartofdredging. com/tcavoidance.htm, both accessed 28 July 2015.

Chapter 11: All the evidence and a song

51. Tinus van Dyk, Johan van den Berg "Unbalanced Wheel" 2002

Chapter 12: Green Gandhi

52. Gandhi reached adulthood during what could have been the perfect storm, and somehow averted it for India. Born in 1869, he reached his prime just as immense bloodshed was spreading across the globe, wrought by unusually violent civil wars, uprisings against colonial rule, genocides, revolutions and political purges.

The twentieth century broke over a lingering colonial conflict in South Africa, where Britain was extending its rule over the country to another region. In the Boer War, guerrilla warfare was being perfected and civilians were imprisoned in concentration camps – another new trend. In addition to some 400 000 horses, more than 25 000 women, children and elderly persons died in these concentration camps.

Just across the border in the present Namibia, Germany had similar difficulty in convincing the population to continue accepting colonial rule.

Early in 1904, less than two years after the end of the Boer War, the Herero people rebelled. After a short conventional battle, survivors were driven into the desert, prevented from reaching water holes and decimated. Perhaps as many as 50 000 people died.

Just two years later in the east of South Africa, the Zulu people rebelled against the British. Again, there were a few conventional skirmishes, again the spear-bearing rebels were surrounded – not by those guarding the exits to the desert but by cannons and machine guns. More than 3 000 Zulus were killed, even more were flogged.

There were many other uprisings, matching in ferocity and losses: The Maji Maji war in Tanzania lasted from 1905 to 1907; Malawi suffered the same fate in 1914/1915; South Africa saw another rebellion in 1914; Madagascar in 1947/1948; and Algeria saw immense bloodshed from

1954 to 1962. These countries all had tiny populations compared to India – and in some cases hundreds of thousands perished.

In addition to the colonial disputes, there were the civil wars of the late nineteenth and early twentieth century. The American version killed about half a million around the time of Gandhi's birth; the Spanish about the same just before the Second World War. Again, these were countries with populations far smaller than India's.

To complete the venomous cocktail there was the phenomenon of conflict arising on the back of communist revolutions, leading sometimes to genocide, starvation, and political purges. The Russian Revolution of 1917 killed some 5 million people. It led to mass starvation in the Ukraine of more than 2 million, and political purges killing at least ten million. In China, in just four years of the Great Leap Forward, more than 2 million people were beaten or tortured to death, while some twenty million starved.

India was subject to the colonial power of Britain. It had a very large population in the same order as China. Furthermore, it had a very stratified population with significant poverty and was ripe for a communist revolution. In addition, it had the explosive element of religious friction created by a society divided between Hinduism and Islam. This element led to significant conflict later in the twentieth century but had been absent from the disputes which killed so many people in places where just some of the other elements were present.

If immense bloodshed was to be avoided, great wisdom and a new approach were was needed. And a very special leader with impeccable integrity.

Along came Gandhi.

As a young legal representative, he lived through all the above-mentioned violent rebellions in southern Africa, and soon saw the futility of armed resistance – on occasion first-hand, as an ambulance driver and medical assistant. But his ultimate philosophy of non-cooperation and passive resistance had a moral foundation: As a student in Britain he'd started to expand his vegetarianism and respect for all life into a philosophy of non-violence and a refusal to do physical harm to friend or foe. Truth, too, was a central tenet, and over time he adapted "God is Truth" to "Truth is God". He would not lie, he would not act violently, but he would refuse to co-operate. For this he was jailed more than once.

His life was one of continuous activism, campaigns, and initiatives. To uplift the Untouchables. To liberate women. To foster respect towards animals. More than anything else, to lead the way in building unity in India, in promoting religious tolerance and in acquiring political independence. By the time Gandhi, now in his late seventies, was assassinated by an extremist, India was independent of Britain; the role of the Untouchables was changing; a line of female political leaders was emerging; and far greater bloodshed in the secession of Pakistan from India (which he opposed) had been averted.

India had dodged the perfect storm, and a new philosophy had been formulated that would influence Martin Luther King, Nelson Mandela and Barack Obama.

53. Folk-rock band Traveling Broke and Out of Gas wrote and recorded a song entitled "Julia Butterfly" released on their 2010 album "CAN YOU Spare Some CHANGE?"

54. Gandhi said: "The greatness of a nation and its moral progress can be judged by the way in which its animals are treated. I hold that the more helpless a creature, the more entitled it is to protection by man from the cruelty of man. I feel that spiritual progress does demand at some stage that we should cease to kill our fellow creatures for the satisfaction of our bodily wants. I do not regard flesh-food as necessary for us at any stage and under any clime in which it is possible for human beings ordinarily to live. I hold flesh-food to be unsuited to our species". See http://www.rrrina.com/wordsofwisdom.htm, accessed 28 July 2015.

55. In an illuminating essay by eco-feminist Marti Kheel, discussing the actions of the Animal Liberation Front (ALF), she describes the values of Green Gandhi:

"Some forms of direct action can be valuable tools for social change. The ALF theft of videotapes from the Head Injury Clinic at the Neurological Institute in Pennsylvania, for example, documented extreme cruelty to primates and was instrumental in closing the laboratory. But when direct action is endorsed uncritically in any and all forms ... it can do more harm than good. There is a marked distinction between liberating an animal from a laboratory, in which no risk to others exists, and burning down a building where someone could be injured."

She goes on to quote, with the necessary approval, Kim Stallwood's criteria for direct action. It should be:

- Motivated by a sense of compassion for all beings, human and non-human;
- tell the truth about the abuse being fought;
- accomplished with an adherence to non-violent principles to all beings (human and non-human) and property;
- undertaken after considering all consequences on all beings; and it should be freely admitted afterwards with full acceptance of the consequences.

See also S. Best and A.J. Nocella (eds), Igniting a Revolution: Voices in Defense of the Earth (AK Press, Oakland: CA, 2006), p 306

56. Having discussed Green Gandhi in this chapter, we have to ask whether it can resolve our PECCAs. Green Gandhi in its very definition allows for a minimum of competition and concentrates almost all its efforts on collaboration. By saying "we will never use violence; we will always admit what we have done; we will always be willing to face the consequences", Green Gandhi becomes a form of gentle, inventive, sometimes even humorous persuasion. Indeed, when an eco-feminist like Marti Kheel argues with considerable sophistication that "the worldview that underlies animal abuse in Western culture derives from a similar mentality to the one which underlies the abuse of women", it is easy to agree.

This eco-feminist perspective eloquently places our environmental challenges and crises as part of a general pattern of oppression and violence. Man oppresses all weaker creatures including woman, who has an innate understanding for the subtleties of the many interlocking systems that make up our natural system. In the process man becomes disconnected from nature. The violence of men spills over into violence against nature.

What the eco-feminists are saying, paraphrased, is: "If ever we should use violence we would already have lost the war – because violence and a violent mind-set is the quintessential problem." This point of view is one that will find agreement from many people – spanning several religions and philosophies. If we agree on the point about the abuse of nature being similar to the abuse of women and the abuse of the helpless, the question then becomes whether, to these activists, the environment is just one battle in a much larger war. It seems that it is.

As for PECCAs, as we will see in subsequent chapters, there is a limited timeframe within which we have to resolve them if we are to survive. It may be that for Green Gandhi, losing the earth is preferable to using violence. And because of this, in the ultimate analysis, a symbolic, non-violent victory is acceptable even if it costs us the earth.

The paradox of competition and collaboration is never accepted in Green Gandhi. Depending on whether you talk to an eco-feminist, to a priest or to Albert Camus, it is always resisted, rejected or scorned. And again we end with the supreme irony of the pacifist Gandhi, extinguished by an assassin.

Derrick Jensen writes:

> "I've always wanted to blow up dams in order to save salmon, sturgeon, and other creatures whose lives depend on wild and living rivers. But that's not right. We need to blow up dams for the rivers themselves, so they can be again the rivers they once were, forever, the rivers they still want to be, the rivers they themselves are struggling and fighting to once again become."

See Best and Nocella (eds), *Igniting a Revolution*, p 287.

Green Gandhi can save a very old and majestic Redwood tree. It can win many symbolic battles, but PECCAs are harder. If there are bigger fish to fry than solving a PECCA, the PECCA is unlikely to get solved. Solving it would be almost incidental. PECCAs can be solved by people who believe that "there is no Planet B". See President Figueres of Costa Rica, at http://www.huffingtonpost.com/andrew-winston/there-is-no-planet-b-and_b_550268.html, accessed 15 June 2012.

Leslie James Pickering says: "The primary responsibility of the struggle isn't to be saintly and righteous, it's to win ... enough is enough". See Best and Nocella (eds), Igniting a Revolution, pp 304–305

57. For quotes from Gandhi see http://www.forbes.com/sites/ashoka/2012/10/02/12-great-quotes-from-gandhi-on-his-birthday/#6f52099b73b6; http://www.brainyquote.com/quotes/authors/m/mahatma_gandhi.html and http://www.mkgandhi.org/nonviolence/phil8.htm, all accessed 23 January 2016;

58. Gandhi. See http://www.mkgandhi.org/nonviolence/phil8.htm, accessed 28 July 2015.

Chapter 14: Blackbeard's Ghost – Disney's accidental master metaphor for human irrationality

59. See Randy Shilts, "And the Band Played on: Politics, People and the AIDS Epidemic" (St Martin's Press, New York City, 1987).

60. Defence mechanisms form part of the Freudian approach that has been more generally accepted than other parts of Freudian theory. There is no unanimity on the exact number of defence mechanisms, but there are probably more than ten. The more relevant ones for present purposes are denial, repression, projection and rationalization. See D.L Paulhus, B. Fridhandler and S. Hayes, "Psychological Defence: Contemporary Theory and Research", in S. Briggs, R.G. Hogan and J.W. Johnson, *Handbook of Personality Psychology* (Academic Press, Boston, 1997), pp 543–579. Also available online, San Diego Academic Press, 1997.

Other defence mechanisms frequently mentioned are regression, acting out, dissociation, reaction formation, displacement, intellectualization, sublimation and compensation. For more information see http://psychcentral.com/lib/2007/15-common-defence-mechanisms/all/1/, accessed 19 June 2012.

61. See Jonathan Baron, *Thinking and Deciding* (Cambridge University Press, Cambridge, 2008), p 224. See also the excellent and accessibly written article by R.S. Nickerson, "Confirmation Bias: A Ubiquitous Phenomenon in Many Guises", *Review of General Psychology*, Vol. 2, No. 2, 1998, pp 175–220, where he says on p 177: "People tend to seek information that they consider supportive of favored hypotheses or existing beliefs and to interpret information in ways that are partial to those hypotheses or beliefs. Conversely, they tend not to seek and perhaps even to avoid information that would be considered counter-indicative with respect to those hypotheses or beliefs and supportive of alternative possibilities."

62. Miguel Cervantes, *Don Quixote*, translated by Edith Grossman (Vintage Books, New York, 2005) pp 58–59.

63. So for instance, the example of the so-called "draw a person" test, which was once believed to be a good psychological diagnostics test. A subject is given a pencil and asked to draw a person. The theory was that the person who draws "projects" her/himself into the drawing. In time, a flawed theory was formulated that paranoid people tended to draw a person with atypical eyes ... that drawing atypical eyes is a sign of paranoia.

    These tests continued to be used even after it had been clearly demonstrated that the assumptions about their usefulness were completely inaccurate. Upon being confronted with this, one clinical psychologist said: "I know that paranoids don't seem to draw big eyes in research labs, but they sure do in my office". It is clear that humans do indeed look for information that will support what they already believe. Perhaps the most ironic example of confirmation bias occurred in the very investigation of the phenomenon and was perpetrated by pioneers in the field. Originally, the accepted wisdom as espoused by developmental psychology experts like Piaget was that people are generally capable of logical reasoning.

By 1960, Watson was testing the underlying assumptions in Piaget's theories. Watson famously gave people the three numbers "2, 4, 6" and asked them to find the "rule" that would describe the common denominator of the figures. The way it worked, the respondents taking part could give a series of three other numbers and ask if they complied with the same rule.

What was observed was that respondents formulated a theory and then gave three figures that fitted their theory. For instance, if they thought the rule was "even numbers ascending" they might say "4, 6, 8" in the hope of a positive answer. If they thought the rule was "numbers two apart and ascending" they might say "3, 5, 7".

By contrast, once a theory explains all that is known, the scientific method operates by falsification. In other words, if I have a theory, I must be able to propose an experiment that could falsify my theory. For example, if I believe that gravity becomes less pronounced in the presence of a highly charged debate, I would propose measuring how long it takes my car keys to fall from my right ear to the ground and then measure it again while I'm having an argument with someone. My theory would be falsified if during the argument the car keys took the same time to fall to the ground as they took to fall in the absence of the argument.

In Watson's study people didn't try to falsify. If they thought the rule was even numbers ascending, they didn't propose the numbers "2, 3, 4" in the hope of hearing "No that's wrong" (which would strengthen the theory because the theory predicts that the answer should be "no"). They always asked questions for which they hoped the answer would be "Yes that's right".

While the reasons for confirmation bias are still being debated, in the world of logic the approach of the subjects that Watson was testing wasn't deemed logical. Rather than conclude that people are not as logical as was initially thought, he and other researchers kept looking for reasons why the experiments might not be displaying the innate logic they were searching for and that Piaget had assumed. The internationally renowned scientists investigating confirmation bias also fell prey to the very same phenomenon. And the band played on, somewhat less disastrously than in the HIV debate.

64. See Nickerson, "Confirmation Bias", pp 193–194 for a stimulating discussion.

65. when physicist Robert Millikan published his experimental work on determining the electric charge of a single electron, he reported only 58 of his 107 observations, omitting from publication those that didn't fit his hypothesis. His notes as he progressed allegedly show comments to results like "Best yet — Beauty — Publish." He won the Nobel Prize in physics for this work.

66. See Drew Westen, Pavel S. Blagov, Keith Harenski, Clint Kilts and Stephan Hamann, "Neural Bases of Motivated Reasoning: An fMRI Study of Emotional Constraints on Partisan Political Judgment in the 2004 US Presidential Election", Journal of Cognitive Neuroscience, Vol. 18, No. 11, 2006, pp 1947–1958. In the 2004 presidential election in the USA, a group of subjects were chosen who felt strong sup-port for either one of the candidates. In the experiment, they were exposed to multiple statements by their favoured candidate; multiple statements from the opposition candidate; and finally, multiple statements from a more neutral third person. The statements in each case seemed to be contradictory but the subjects were then given explanations for the respective contradictions ... contradictions that had the potential to remove the inconsistency. The subjects were then asked to assess which of the three persons had made contradictory statements.

The authors begin their article by remarking that "people are rational within limits imposed by cognitive shortcuts and heuristics" and then go on to say:

In political science, a long-standing body of research on ''partisan'' biases in political judgment (e.g. Taber, Lodge, & Glathar, 2001; Campbell & Converse, 1960) points to another set of limits to rational judgment imposed by emotion-biased or motivated reasoning ... Motivated reasoning can be viewed as a form of implicit affect regulation in which the brain converges on solutions that minimize negative and maximize positive affect states ... Freud (1933) described such processes decades ago, using the term "defence" to denote the processes by which people can adjust their cognitions to avoid aversive feelings such as anxiety and guilt.

In the experiment, perhaps not surprisingly, the subjects overwhelmingly said that the opposition candidate had made contradictory statements. The subjects were assessed while being monitored with an MRI scanner. What is utterly remarkable is that especially when assessing the contradictions of their favoured candidate, emotional loci in the brain that have nothing to do with reasoning were activated. The conclusion drawn from the experiment was that the subjects were actively (and emotionally) blocking their own recognition of their candidate's inconsistencies.

67. See http://en.wikipedia.org/wiki/Censorship_of_images_in_the_Soviet_Union, accessed 28 July 2015.

68. A study was conducted on people who either believed or did not believe in extra sensory perception (ESP). They were informed about the results of experiments on the subject. The believers and non-believers were divided into two further groups, of which one half were told the study supported the existence of ESP while the other half were told it did not. All four groups were then asked to recall the material. The recall was good in the entire non-believer group and good in the believer group who'd been told the evidence supported ESP. However, in the believer group who'd been told that the evidence did not support ESP, the recall was poor. They remembered less. In the case of some, their memories had twisted the evidence around and they recalled the evidence as having not opposed but supported the existence of ESP. See Dan Russell and Warren H. Jones, "When Superstition Fails: Reactions to Disconfirmation of Paranormal Beliefs", *Personality and Social Psychology Bulletin*, Vol. 6, No. 1, 1980, pp 83–88.

69. The implications of the above are really quite mind-blowing. Let's assume for the moment that most people want a life that:

- Is happy
- Is prosperous
- Is sustainable
- Is somewhat green
- Just happens without disruptions or sacrifice
- Will be the same or better for their children and for their grandchildren.

If this is assumed then it must also be assumed that collectively we will and must have a gigantic environmental blind spot and that we will be unable to see anything that threatens the above. A PECCA may thus be invisible in principle to the majority of people. The more important and threatening it is, the more invisible it will become. This is why green scare-mongering is so counter-productive.

70. A fascinating case study of full circle, occasional rationality in US environmental policy is the treatment of oil rigs and renewable energy in terms of the safety studies that have to be carried out before the activity can be commenced.

In the late 1960s, a few things happened that changed the landscape of environmental regulation in the United States. Perhaps foremost there was so-called flower power, with the hippy era at its pinnacle. There were many songs about nature. If you were going to San Francisco, you had to be sure to wear some flowers in your hair. Carol Kind sang "You make me feel ... like a natural woman." Joni Mitchell was singing "Big yellow taxi". In just the previous few years, there had been "Green, green grass of home" by Tom Jones; "Mother nature's son" by The Beatles; "Flowers in the rain" by The Move; "River deep, mountain high" by Ike and Tina Turner and "Where do the children play?" by Cat Stevens. In 1969, the stratosphere-bound "Earth wind and fire" was founded by Maurice White.

Also in 1968 was the disastrous Santa Barbara oil spill. Late in January, on an off-shore oil rig, there was a "blow-out". In the following days, between 80,000 and 100,000 barrels of crude oil spilled into the ocean and onto the Californian beaches over a wide area. Thousands of sea birds, dolphins, and other marine life species were killed.

The impact of Rachel Carson's book about environmental damage called *Silent Spring* (1962) was still quite strong at the time.

The result was that a number of new federal laws were enacted to strengthen environmental protection. These included the Wilderness Act; the Clean Air Act; and the Clean Water Act.

Within this context the National Environmental Policy Act of 1969 was born. Along with its other provisions, it contains the framework for the investigation of environmental impacts. Over time the legislation developed to the point where three levels of investigation were identified:

> In cases where serious investigation was required, an Environmental Impact Statement (EIS) needed to be prepared.
>
> In cases where less investigation was required, an Environmental Assessment (EA) had to be complied and a finding of No Significant Impact (FONSI) would be made.
>
> In cases where practically no investigation was required, a Categorical Exclusion (CE) could be provided by the relevant authorities – after which the proposed activity could proceed.

This latter case refers to "a category of actions which do not individually or cumulatively have a significant effect on the human environment and which have been found to have no such effect in procedures adopted by a federal agency in implementation of these regulations (section 1507.3) and for which, therefore, neither an environmental assessment nor an environmental impact statement is required." An agency may decide in its procedures or otherwise, to prepare environmental assessments for the reasons stated (in section 1508.9) even though it is not required to do so. Any procedures under this section shall provide for extraordinary circumstances in which a normally excluded action may have a significant environmental effect.

See http://ceq.hss.doe.gov/nepa/regs/ceq/1508.htm, accessed 21 June 2012.

The detailed legal aspects can be debated, but the intent is clear: If you want to do something dangerous, investigate it first. If you're really safe, you don't have to investigate. Somewhere in the middle, do some investigation. "Extraordinary circumstances" typically include such matters as effects to endangered species, protected cultural sites, and wetlands.

When anyone wants to build a wind farm in the USA, they would typically need to discharge the heaviest burden of proof and prepare a full Environmental Impact Statement (EIS). The reason is explained as follows on a government website:

"Federal laws and regulations require the federal government to evaluate the effects of its actions on the environment and to consider alternative courses of action. The National Environmental Policy Act of 1969 (NEPA) specifies when an Environmental Impact Statement (EIS) must be prepared. NEPA requires that an EIS be prepared for major federal actions with potential for significant impact on the quality of the human environment. The BLM has determined that amending land use plans and the establishment of a Wind Energy Development Program would be major federal actions as defined by the NEPA, and, thus, the BLM has prepared an EIS". See http://windeis.anl.gov/eis/why/index.cfm, accessed 21 June 2012.

If recent projects requiring EISs are perused, several wind farms do indeed feature. See http://yosemite.epa.gov/r10/ecocomm.nsf/b9d67f6000e5b58888256e5900642421/b9f3524d0c457af688256e6a0082b5ed!OpenDocument, accessed 21 June 2012.

The conclusion we should draw is that wind farms are amongst the activities for which profound investigation is needed to prevent environmental damage.

Wind farms will always tend to stand unfairly opposed to the "status quo" of fossil fuels. This is due to confirmation bias in general and the primacy effect in particular. So what happens in fossil fuels – and more particularly in off-shore oil drilling? Given that an off-shore oil rig blow-out in Santa Barbara in 1968 was the very reason why the principle of environmental impact assessments was instituted, do these activities also attract the heaviest burden in terms of the level of investigation required, like wind farms do?

Between 1955 and 2007, forty four major oil rig blow-outs occurred worldwide. See http://en.wikipedia.org/wiki/Blowout_(well_drilling), accessed 21 June 2012. This equates to a major blow-out every 14 months or so. Fatalities occurred often, with 22 people killed in 2007 alone. And yet, by 2007, oil was probably more valuable than gold dust. In the Gulf of Mexico, new but complex technology had been developed to drill far, far deeper than had been the case at Santa Barbara, and on a far broader scale. In 2009, quite amazingly, the authorities considered the Deepwater Horizon drilling operation and granted it a "categorical exclusion" from all environmental reviews, along with hundreds of similar exclusions being granted to other drilling operations in the Gulf.

This implies that in the opinion of the authorities, drilling for oil at the most extreme depths ever attempted constituted:

> "a category of actions which do not individually or cumulatively have a significant effect on the human environment ... and for which, therefore, neither an environmental assessment nor an environmental impact statement is required."

When the Deepwater Horizon blew out a year later it killed fewer people than the 2007 incident, but 11 people still lost their lives. The lawsuits emanating from the spill is projected to be nearing settlement for an estimated US$7–9 billion.

No rational explanation seems imaginable of how it was decided that the Deepwater Horizon posed less danger than a few wind turbines on a hill. And yet, the absence of a rational explanation doesn't preclude the presence of an irrational explanation. Blackbeard's Ghost.

71. In a ground-breaking experiment, people who had strong views for or against capital punishment were chosen. They were then each given

two reports dealing with whether capital punishment was a general deterrent to violent crime (murders committed decreased when capital punishment was instituted). In the one report, the clear conclusion reached was that capital punishment was a deterrent to violent crime. The other report reached a clear conclusion that capital punishment was not an effective deterrent to violent crime. Both reports were sophisticated fabrications by the researchers.

The participants read both reports. Both proponents and opponents of the death penalty were influenced more by the report supporting their pre-existing belief than by the other report (which did not support their pre-existing belief). They thought the report that agreed with them was "more convincing" and that the other report was "weak". The result of the mixed evidence was polarisation: Both camps believed in the same as before, more strongly than before. The experiment had served to drive them further apart.

Jonathan Baron of the University of Pennsylvania, author of highly regarded book *Thinking and Deciding* (Cambridge University Press, Cambridge, 1988) concludes on p 210: "This polarization study is disturbing because it suggests that evidence is useless in settling controversial social questions."

This exact dynamic seems to be playing out in Climate Change. There more evidence emerges, the stronger denialism becomes. See "The science for climate change only feeds the denial: how do you beat that?" at https://theconversation.com/the-science-for-climate-change-only-feeds-the-denial-how-do-you-beat-that-52813, accessed 1 February 2016

72. The late Douglas Adams coined the term "Somebody Else's Problem" (SEP):

"An SEP is something we can't see, or don't see, or our brain doesn't let us see, because we think that it's somebody else's problem ... The brain just edits it out, it's like a blind spot. If you look at it directly you won't see it unless you know precisely what it is. Your only hope is to catch it by surprise out of the corner of your eye.

The technology involved in making something properly invisible is so mind-bogglingly complex that 999,999,999 times out of a billion it's simpler just to take the thing away and do without it ... The 'Somebody Else's Problem' field is much simpler, more effective, and can be run for over a hundred years on a single torch battery. This is because it relies on people's natural predisposition not to see anything they don't want to, weren't expecting, or can't explain". See Douglas Adams, *Life, the Universe and Everything* (Pan Books, London, 1982), p 26 and further.

73. Confirmation Bias has an interesting effect on the so-called "precautionary principle".

In the last 20 years, the precautionary principle has become ever more prevalent in conversations, treaties and laws about the environment. There are different definitions to the precautionary principle, but essentially it boils down to who bears the onus of proof. In a criminal law case, the State ("Prosecution"/"People") normally bears the burden of proving the case. If there is insufficient evidence, the status quo persists as it existed before the trial: The accused "walks" – is acquitted. If the evidence is sufficient the accused is convicted and punished – the status quo changes. In civil court cases like contractual disputes, the "onus" to prove the case or the critical aspect of the case is normally on the one making the allegation. If I claim breach of contract, I must prove it. If I claim you have slandered me, I must prove it. If I fail to bring sufficient evidence, the case is dismissed and the status quo remains. If I bring sufficient evidence, a judgment is given in my favour and the status quo changes: You are ordered to compensate me.

In environmental matters, there is often scientific uncertainty: Will this new housing development really cause the extinction of a rare species of owl? The developer will ask for permission to build. An objector who cares about owls will ask that the development not be allowed because a rare owl species is likely to become extinct. By normal principles, unless the objector brings sufficient evidence that the owl species will indeed become extinct, the objection will be dismissed and the development be allowed to proceed.

The precautionary rule reverses this and says that in the case of uncertainty, it is for the developer to show that harm will not occur and that if he produces insufficient evidence, he will not be allowed to proceed because it is "better to be safe than sorry".

At the Earth Summit in 1992, the Rio Declaration was formulated. Principle 15 states:

"In order to protect the environment, the precautionary approach shall be widely applied by States according to their capabilities. Where there are threats of serious or irreversible damage, lack of full scientific certainty shall not be used as a reason for postponing cost-effective measures to prevent environmental degradation." See http://www.unep.org/Documents.multilingual/Default.asp?DocumentID=78&ArticleID=1163, accessed 11 October 2012.

The precautionary principle has been accepted into EU law, being reflected in article 191 of the Lisbon Treaty. It has also entered legal systems in the USA, Japan, South Africa and Australia, to mention just a few.

The precautionary principle is very important to some of the more prominent PECCAs like climate change and biodiversity. If we had to follow it, we would say: "irrespective of whether it's scientifically agreed exactly what damage would occur under what circumstances, we are going to take steps to avoid that damage".

This would make the Policy (P) and Collective Action (CA) components of these PECCAs somewhat easier – at least we would no longer have to argue about whether these issues should be dealt with or not. The stark truths emanating from our investigation into Confirmation Bias show us why the precautionary principle hasn't made much headway. The simple reason is that the precautionary principle asks us to do something that is against human nature.

Confirmation Bias holds that: "... the status quo is good and will persist. Every time we get new evidence about its welfare, we'll see if it confirms that the status quo is sustainable. If yes, we'll remember it and feel ever more confident about our future. If no, we'll ignore the evidence. If somehow we can't ignore the evidence, we'll forget it. If we can't forget it, we'll remember it as having been supportive of the sustainability of everything. And given how wonderfully things are going, anyone who wants to throw money at changing the status quo has to be insane".

74. By way of example, there is a strong unanimity between mainstream scientists that anthropogenic or "man-made" climate change is now proven. This does not prevent an array of Blackbeardesque minority views on the subject, which can be summed up as follows.

Minority approach 1: Other reasons

All observed climate change is due to natural factors and is not impacted by people. So, for instance, higher temperatures would be due to increased solar activity – which occurs in cycles. When the solar cycle changes again, the temperature will drop again. Everything moves in cycles (like ice ages and warmer periods in the past). See, for instance the following passage, which (while not being used for the purpose by the authors) mentions the facts sometimes used by others to justify the "other causes" theory:

"Since the earth formed more than four billion years ago, its climate has periodically shifted from warm to cool and back again – sometimes dramatically ... About 8,000 years ago Saharan North Africa, now an arid desert, was home to numerous wetlands and lakes dotted shoreline human settlements ... Only 1,100 years ago Viking settlers took advantage of a particularly mild and warm period to colonize a temperate area that they named "Greenland"... at the same time, the great Mayan civilization of central America collapsed. Climate change and prolonged drought is one of the theories attempting to explain the sudden and mysterious collapse of the Mayan civilization. See John T. Hardy, *Climate Change: Causes, Effects and Solutions* (John Wiley & Sons, Hoboken: NJ, 2006), p 23.

Minority approach 2: Scientific agnosticism

"There is more than one school of thought. The science is uncertain and the existence of climate change remains unproven. Until such time as we know for sure, we should do nothing". A variation of this is to acknowledge a human impact on the climate but to state that "it is impossible to know how much is natural and how much is man-made" and that until we know, we should not act. See the Union of Concerned Scientists, at http://www.ucsusa.org/global_warming/science_and_impacts/science/certainty-vs-uncertainty.html, accessed 13 May 2012.

Minority approach 3: Evidence rejection

"There is no evidence of climate change". See http://www.globalwarminghysteria.com/ten-myths-of-global-warming/, accessed 13 May 2012.

Minority approach 4: Conspiracy

"There is a global conspiracy to misinform and mislead the public about the real facts in order to maintain a certain social order. In fact nothing is happening – humans are not affecting the climate. The evidence has been forged." On this see *The Greatest Hoax: How the Global Warming Conspiracy Threatens Your Future* by (US senator) James Inhofe (WND Books, Washington, 2012). WND Books has published works on a number of conspiracy theories, including *The Late Great USA: The Coming Merger with Mexico and Canada* (2007) by Jerome Corsi; and *Manchurian President: Barack Obama's Ties to Communists, Socialists, and Other Anti-American Extremists* (2010) by Aaron Klein.

See the following comment on *The Greatest Hoax*:

" ... within the pages of *The Greatest Hoax*, Inhofe has marshaled the most compelling evidence against global warming, eviscerating the left's justification for a system of regulations and taxes that would permanently hobble the US energy sector and create a new bureaucracy that would exert, directly or indirectly, control over each and every American's life. The fraud involved and the dangers to freedom and prosperity posed by this movement are so great, that Inhofe describes it as nothing short of 'the greatest hoax ever perpetrated against the American people'" Inhofe is now leading the charge to definitively expose the Greatest Hoax." See http://greatesthoax.com/. The website seems to have been discontinued now.

Minority approach 5: Adaptation

Species will adapt to a changing climate and we don't have to worry about it." A variation is "new species are evolving as quickly as old ones are going extinct".

Minority approach 6: Good news

"If the earth becomes warmer, it will be good for us. We'll have longer growing seasons and our crops will be better." On this see http://www.conservapedia.com/Global_warming#Beneficial_effects_of_Global_Warming, accessed 28 July 2015; and the following excerpt at the same website:

> "Some researchers point out that benefits of health global warming have been overlooked, or minimized. As far back as 1996, Thomas Gale Moore, Senior Fellow at Hoover Institution (Stanford University) contended that positive health and amenity effects would be a result of projected increases in temperature. Another source notes, in areas that see extreme cold temperatures, deaths related to colder weather would drop significantly, leading to decreased health care costs. Warmer temperatures would also mean less energy use to heat homes and buildings, helping to conserve energy. With the changes brought about by global warming more land would become available for uses like farming and living. Forests and plants would grow stronger, healthier, and more abundant because of the warmer weather, and this would mean more oxygen being released into the atmosphere."

Minority approach 7: Deliverance

"Even if climate change turns into a very big problem, we will be delivered from it". This could occur, depending on the perspective of the person, through divine intervention, technology (e.g. the resettlement of humans in space) or through a transformation of human consciousness to a higher level never attained before. See, for example, "Only God is the answer and key to the problems of global warming. Acknowledge His existence and omni-importance and He will deliver us from this global scourge." See at https://www.newsday.co.zw/2010/11/16/2010-11-16-christians-have-role-in-fight-against-climate-change/; and "If we could move a part of the human population to space, the pollution and greenhouse gas emission would decrease and thus the global warming problem would be resolved. This is why space colonization is so important to our home planet." See also http://www.cosmosfrontier.com/space-colonization/dangers/global-warming.html; and "... the inhabitants of Mars and Venus destroyed their respective planetary homes because they raised too much livestock, and the gases released triggered an irreversible greenhouse-gas effect, plus poisonous hydrogen sulphide in the case of Mars. So, that's why we see only the traces of the landforms and oceans that once used to be there. Around five-billion planets have been destroyed or have met a similar fate as Venus or Mars. Only those planets that have inhabitants with a low level of consciousness have been destroyed or ruined, not the high level ones." See http://crisis2peace.org/book.php?b=&wr_id=45, websites on the deliverance approach accessed 13 May 2012.

Chapter 16 The dual DNA of the World

75. The story of the parts corresponds roughly to *reductionism*, which can mean two slightly different things:

Generally, regarding complex things, that they can be understood by breaking them into parts and interactions. (For example, a house consists of the bricks, mortar, ceilings, wiring, etc.)

More specifically, when trying to understand a complex system, that if the parts are known and understood the system itself is known and understood (For example, if you read every chapter of this book, you have read the totality of what the author is trying to convey). See: Encyclopedia Brittanica online at http://www.britannica.com/EBchecked/topic/494866/reductionism; and the Internet Encyclopedia of Philosophy, at http://www.iep.utm.edu/red-ism/, both accessed 13 May 2012.

Newton's particle theory is an example of reductionism. See http://pespmc1.vub.ac.be/NEWTONWV.html, (accessed 13 May 2012). It is the example of the house applied to the universe – except it adds that given the dynamic nature of matter, it will behave to a large extent like billiard balls. Collisions will create predictable results. One conclusion that could be drawn from Newton's laws is that everything can be understood and predicted if sufficient knowledge can be gathered. This implies that if I know the position and speed of every billiard ball on the table then the future of the table and everything on it is known exactly. Knowledge is power – and knowledge is understanding all the small parts that make up the entity.

For a long time, many people felt that not everything could be explained this way. For instance, the philosopher Hegel said that a substance like blood is an organic unity and cannot be understood by breaking it down into the constituent parts/chemicals that form blood. The chemicals that make up blood, if put together, do not make blood. There was, however, no apt term for the belief that some things should be understood in terms of their entirety rather than in terms of their parts.

76. The story of the whole corresponds roughly to the theory of holism as it ultimately manifests in the Gaia Theory. Someone who instinctively understood this concept at a time when no name existed for it, was Eugene Marais, the extraordinary South African natural scientist, lawyer and poet. Marais began studying as a medical doctor in London in the last years of the nineteenth century, before turning to legal studies at Inner Temple and going on to become a barrister. From 1910, at the age of 39, he retreated to his farm north of Pretoria where he began studying various species like baboons, termites, and snakes. He became the first man to conduct a prolonged study of primates in the wild and recounted that he had lived amongst the baboons for three years. Marais wrote books that were (posthumously) translated into English as *My Friends the Baboons* (1939) and *The Soul of the Ape* (1969). It was his work on termites that was the most ground-breaking and controversial. His intensive field studies led to a series of articles in a popular magazine in which he set out his theory on termites, later translated as "The Soul of the White Ant" (1937).

Marais' theory states that a termite colony should be considered as a single organism. By way of comparison to a human, the queen was the equivalent of the brain and the womb; the workers were mouthparts and tissue builders; the soldiers were the white blood cells; and the humus gardens were the stomach.

Here we observe reductionism in reverse. The inquiry is not "of what does a termite consist?" but rather "Of what is a termite part?" Marais' theory, more than anything else, shows a predilection, a certain perspective. An ant does consist of something, and Marais could well have spent days and years finding out exactly what the component parts were. But he wanted to know what was hierarchically above and beyond the ant. The mindset that he embodied clearly held an allure.

Within three years after Marais' articles, Jan Smuts' book *Holism and Evolution* (1926) was published. Smuts defined holism as: "The tendency in nature to form wholes that are greater than the sum of the parts through creative evolution".

Albert Einstein read Smuts' book. He was intrigued, and wrote to Smuts predicting that the latter's mental construct of holism would (together with Einstein's own work on relativity) "direct human thinking in the next millennium". Indeed, it seems Einstein himself must have had this same holistic mind-set in formulating his theory of special relativity with its inter-relational aspects that for instance allowed time (already broken down by others to years, months, hours, minutes, seconds) to be conceived as being part of something larger – the space/time continuum.

Suddenly, the term "holism" was there to describe the mind-set of Marais, Smuts, Einstein and many others going back, for example, to Baruch Spinoza in the mid-seventeenth century. Around this term discussions could take place and disciplines could grow. Whether Einstein was correct about the profundity of the holism paradigm is as yet unclear, but in the intervening eight decades it has led to notable progress in the fields of systems thinking; complexity theory; as well as gestalt psychology; teleological psychology; economics and especially in the emerging discipline of ecology.

77. See the discussion on the Penn State Geosciences website at http://www3.geosc.psu.edu/~dmb53/DaveSTELLA/Daisyworld/daisyworld_model.htm.

78. While he was working for NASA in the 1960s, James Lovelock formulated the Gaia Theory. He named it after the Greek goddess Gaia, who in mythology was the "earth mother" of all. Eugene Marais had looked at the termite colony and had not asked "of what does the termite consist?" but "of what is the termite part?" This had led him to the theory that a termite colony was a single organism.

In the same way Lovelock looked at non-living matter, bacteria, protista, fungi, plants, animals and all human endeavour, including the planet on which it all occurs, and said that it all forms one complex, interacting system that can be thought of as a single organism. Furthermore, this "biosphere" in our own case has a regulatory effect on the Earth's environment that acts to sustain life (Daisyworld was conceived to demonstrate the theory mathematically).

One of the attractions of the theory has been its name: just like the coining of the word "holism" had enabled further discussion and investigation, the arresting concept of the entire earth and all its beings as a single "mother" has created a potent (if highly controversial) metaphor to easily communicate and grasp the notion that the earth is a single being.

There is good evidence of certain critical variables that have been very stable for a long time, despite occurrences that have upset the balance. Ocean salinity has been constant around 3.4%. On this see *Gaia in Action*, edited by Peter Bunyard (Floris Books, Edinburgh, 1996) p 99. Furthermore, the atmospheric composition has been very constant at 78.09% nitrogen, 20.95% oxygen, 0.93% argon, and 0.039% carbon dioxide. See James Lovelock, *The Vanishing Face of Gaia* (Basic Books, New York City, 2009), p 179. The theory states that it is the very presence of life that has caused this stability – just as in Daisyworld.

Nearly fifty years have passed since the theory was formulated, and more than thirty since it was widely published and discussed for the first time. It has received support in fields such as bio-diversity, where experiments show that the health of all organisms in an eco-system improves as more numbers are added. Daisyworld predicted this and in a later version it has grey (in-between) daisies that further stabilized the planet to the benefit of all daisies.

In other fields of science, the Gaia Theory has been vigorously contested. Various conferences have been held and a number of nuances of the theory developed, ranging from "weak Gaia" (living organisms and the planet influence each other without necessarily leading to homeostatis) to "strong Gaia" (the earth is a living being with consciousness that consciously manipulates conditions to the benefit of life generally). As a general principle, reductionism and holism seems to attract proponents that are temperamentally different to the point of polarization.

However, Juan-Franco's juxtaposition, positing that the world is both – sometimes the one and sometimes the other – is not a view widely held at this time

Chapter 18: Bruce Willis and Green Die Hard

79. "We must bury the dead, and feed the living". These wise and pragmatic words were reportedly uttered by the Marquês de Alorna after the great earthquake of Lisbon on 1 November 1755. More information available at http://www.lonelyplanet.com/portugal/lisbon/history, accessed 28 July 2015.

80. Opposition can quickly turn violent when there is a breakdown of trust and a sense that nothing else helps. In the Pacific northwest of the United States, the authorities decided to fell more than 40 beautiful, old trees to make way for a parking lot. Many people objected and a public meeting was scheduled to discuss the matter. The city reacted by quietly planning to bring the tree felling forward with the result that by the time the meeting takes place, the trees will be gone. Trying to stave off the cutting of the trees until the meeting the following day, activists climbed into the trees. The city reacted by sending law enforcement to remove them, cutting open their clothing and spraying their genitals and eyes with pepper spray. Somewhat later that day, they were forcibly removed and the trees cut down. The breach of trust unleashed

a dramatic escalation. End of discussion. End of the trees. These events are captured in Marshall Curry's Academy Award Nominee documentary "If a Tree Falls: A Story of the Earth Liberation Front". The film is well balanced yet disturbing. It shows the young people who graduated from peaceful protest to arson and property destruction. They are interviewed and shown at their homes, doing everyday things. So are the law enforcement officers who prosecuted them and the logging company owners who suffered at their hands

The adherents feel that action of the direct kind is required. The cyclone is coming. By "direct action" they mean violence – so far not violence that harms humans, but certainly violence that destroys property. These are the Bruce Willises of the more extreme environmental movements. They organise themselves into secret "cells", have clandestine plans and agendas, and generally hide from the law. Most have regular day jobs, doing this and that, and do their "ecotage" under cover of darkness. The next day they are back at work, but someone's gas-guzzling SUV may be no more. Or a laboratory where tests are done on animals may have been raided, the animals released, the facility torched.

In a multi-year operation code-named "Backfire" the FBI "turned" one of the inner circle of the Earth Liberation Front and then flew him around the country wired with a recording device, accidentally "bumping into" other ELF members and soliciting damning statements from them about their complicity in prior acts of sabotage. On this the FBI started building its case. On 7 December 2005, they struck simultaneously in several US states, arresting Daniel McGowan, William Rodgers, Stanislas Meyerhoff, Kendall Tankersley, Chelsea Gerlacha and Kevin Tubbs. Rodgers committed suicide in his cell, while some of the others negotiated lighter sentences by agreeing to testify against their co-accused. Later, Josephine Overaker, Suzanne Savoie, Joseph Dibee, Jonathan Paul and Rebecca Rubin were also arrested.

The accused faced 30 years or more in prison. In the end, the sentences were shorter – McGowan was jailed for 7 years. However, the prosecution was successful in arguing for a "terrorism enhancement" in his case, leading to his being treated as a terrorist and being incarcerated in the extremely harsh and restrictive Communication

Management Unit at the Marion Penitentiary Facility. McGowan was allowed a fifteen-minute telephone call once a week and a four-hour visit through a glass pane once a month. He was never allowed to go outside.

Savoie was incarcerated in a similar facility but she has now been released. At the time of writing McGowan is still incarcerated.

The terrorism enhancement question had effectively put violent environmental activism in the accused's box and determined that in the United States, such actions are classified as terrorism.

While conceding that these organisations generally do not harm humans, Jarboe calls them a "serious terrorist threat" – the FBI estimated at the time that the ALF/ELF had committed more than 600 criminal acts in the United States in the previous six years, resulting in damage in excess of 43 million dollars.

Viewing Curry's film one has the distinct impression that the Earth Liberation Front has been utterly defeated by the machinery of justice. The causes they'd been fighting for seems to have been deserted, at least by the more radical component. In a post 9/11 world, there is no space for a terrorist – or an eco-terrorist. Unlike the original, these John McClanes have backed off.

81. There is usually a "message" being "delivered". If these "eco-terrorists" are burning an SUV they are saying that it is immoral and wrong to drive a vehicle with such low fuel efficiency and willy-nilly put more CO_2 pollution into the global atmosphere. If they are destroying driftnets they say that driftnets are causing excessive over-fishing in our communal, over-fished oceans. If they steal videos from a laboratory to show the media how helpless primates are deliberately inflicted with whiplash so that treatment for humans can be tested, they maintain that humans have no right to test treatments on animals to the benefit of themselves.

82. Best and Nocella (eds), Introduction to *Igniting a Revolution*, note 1, p 25.

83. Best and Nocella (eds), *Igniting a Revolution*, pp 284–287.

84. Leslie James Pickering, in Best and Nocella (eds), *Igniting a Revolution*, pp 301–305.

85. It's in the nature of PECCAs that they don't go away. If they don't get attention, they don't develop inferiority complexes and skulk into corners any more than giant meteorites do – they just keep coming. Eventually they get noticed.

    A single superpower or a few strong countries could in future act unilaterally and with violence in order to secure a safe outcome.

    Let's pose a theoretical situation and then ask ourselves a theoretical question: Suppose all the important scientists and politicians of a superpower agree that global greenhouse gas emissions have to decrease dramatically. TODAY. RIGHT NOW. If they don't, a calamitous tsunami will be created at dawn tomorrow by a giant earthquake created by the extra heat of climate change, and this tsunami will kill tens of millions of the superpower's people and destroy its economy for decades. The superpower is armed with many fighter planes, ships, submarines. Also with nuclear weapons and biological weapons of mass destruction. The rest of the world doesn't believe the scientists – they think it's all scaremongering. They're refusing to lower their emissions and anyway, they cannot do so except by changing technology over the course of many years.

    The crisis is hours away, and something must be done. It is "us or them" time. The superpower, like John McClane, surely isn't just going to lie down and die. It will act with all necessary military might. It might decide to send bombers, to destroy every major refinery outside their own borders. The result would be no crude oil for months, no transport of goods, no vehicles moving, no ships moving. Civil unrest, chaos, civil wars, starvation. They might bomb some key installations that use a lot of energy. They might blow up coal-fired power plants. These typically take ten years to build – that's when a country's cars and ships are moving and the population isn't starving. And then the superpower might still say to itself that it no longer buys the line that a larger world population doesn't necessarily imply more pollution – and unleash its biological weapons. Or even its nuclear weapons.

All of this is more than possible, it is inevitable. Trumping the problem in unilateral fashion. In the scenario we've painted, it is the only possible outcome. The only reason why it cannot happen today is because at present the scenario we painted isn't the reality. The "cyclone" and its point of no return isn't perceived as close enough yet.

PART 3 UGLY LIES

Chapter 20: Artistic murder, rational suicide

86. The Apology above. For additional material in this section the author is indebted to the same source, but the information has been shortened and simplified.

Chapter 22: propaganda: the seesaw's second prejudice

87. In the last sixty years, there have been some bruising policy battles globally on issues such as tobacco smoke (does smoking cause cancer?); acid rain (do sulphur emissions cause acid rain/does acid rain harm the environment?) and even nuclear weapon defence systems (can "we" bomb the other side into oblivion without harming the global environment to an extent that would ensure self-destruction?). In the United States, sound data exists to investigate these battles and learn some lessons from them. In *Merchants of Doubt: How a Handful of Scientists Obscured the Truth on Issues from Tobacco Smoke to Global Warming* (Bloomsbury, London, 2011), Naomi Oreskes and Erik M. Conway look at six decades of scientific/political debates in the USA. The striking discoveries they make are:

    - The same arguments against change were used by proponents of the status quo in fields as diverse as the dangers of tobacco smoke; acid rain; the damage to the ozone layer emanating from CFCs; and the effort to address global climate change – even in strategic missile defence (see chapter 2 on the tobacco disputes; chapter 3 about acid rain; chapter 4 about the ozone layer; and chapter 6 on climate change).

- In each case scientific "doubt" was alleged long after the reputable body of scientists had accepted the facts that required change. Scientists were accused of "scare-mongering" and hysterical over-reaction.
- Contrary explanations for these danger signals were provided that were reassuring and emotionally "easier to believe". These explanations were based on very little or no original research and were seldom supported by any meaningful peer review.
- The very same scientists were quoted as "experts" in fields as diverse as the danger of tobacco smoke; acid rain; the destruction of the ozone layer by man-made chemicals; and in anthropogenic climate change (the implication is that these people, while intelligent, were either "hired guns" willing to reverse engineer science from the conclusion wanted to the evidence that exists – and reconcile the two. Alternatively, they were so blinded by ideological and patriotic fervour that they did the same in the honest belief that it was good science).
- The inclination of the media (sometimes backed up by laws about media access) throughout, is to say that if there is a single divergent opinion then the science is "unsettled" – no matter who disagrees, who funds the disagreement and irrespective of the quality or merit of the divergent opinion. (Known as the "fairness doctrine", this is tantamount to stating in a newspaper report, that if there is one man globally with a PhD saying the world is flat, that the science on whether the earth is flat or round is as yet "unsettled").
- In each case, meaningful action was delayed for a very long time.
- In each case, the peer-reviewed scientific consensus that had existed for many years before meaningful action was taken was eventually proven to be correct and in each case barring climate change (so far), the necessary majority for action was eventually mobilized.
- Implicitly, what Oreskes and Conway are saying is that the precautionary principle is a dead duck when it requires positive action with concomitant societal change. If you want to build a casino on a wetland and you might cause a rare species of riverine rabbit to go extinct, it is easy to invoke the precautionary principle and to order that the development should cease and the status quo should remain. That is the precautionary principle used negatively –

to prevent action. If, however, we need to use less energy, drive smaller cars and take holidays closer to home because (maybe) we would be in trouble otherwise, nothing happens (positive precautionary action – a call to act). In such an instance people generally do not change themselves and their society to avoid a possible damaging outcome. They stick to the status quo.

In general, had it not been for sell-by dates of possible solutions, we might have taken comfort from the trend lines that emerge from Oreskes and Conway's engrossing book. Bogus research and misinformation are sponsored by rich entities wanting to maintain the status quo, but as time goes by, good science and logic and rationality often get the upper hand.

Summary of policy debates in the US over the last six decades (sourced from Oreskes and Conway)

Issue	Peer reviewed scientific conclusion	Time of scientific acceptance	Comments by status quo to scientific consensus	Necessary action taken (when and after how long)
Smoking	Smoking causes cancer	Late 1950s – even tobacco industry scientists accepted this early in the 1960s	The link between smoking and cancer is unproven	After 60 years in 2009 – US Congress gives FDA the authority to regulate tobacco
Acid rain	Pollution from power plants, cars and factories cause acid rain	1979–1981	Not certain that acid rain is being caused by man- made pollution; no need for hysteria; not a big problem	After about 10 years in 1990 – cap and trade scheme introduced that would lower emissions by 54% at negative cost

Ozone layer	Man-made CFCs cause ozone depletion ("a hole in the ozone layer")	1976	... pompous claptrap; ...human impact too small to matter; "...it's all due to volcanoes	After about 12 years, the 1988 Montreal Protocol is accepted in USA and Du Pont decides to phase out CFCs

If we look at the table above, we see that it has taken between ten and sixty years for science to gain the necessary majority to ensure Collective Action (CA). In a very interesting development, Richard Muller, climate sceptic who used to do work on climate change related issues for the denialist Charles J. Koch Foundation (part of the intricate web of non-profit associations mentioned below), has now concluded that the larger body of scientific consensus has been correct. Muller says:

"Call me a converted skeptic. Three years ago I identified problems in previous climate studies that, in my mind, threw doubt on the very existence of global warming. Last year, following an intensive research effort involving a dozen scientists, I concluded that global warming was real and that the prior estimates of the rate of warming were correct. I'm now going a step further: Humans are almost entirely the cause ... Our results show that the average temperature of the earth's land has risen by two and a half degrees Fahrenheit over the past 250 years, including an increase of one and a half degrees over the most recent 50 years. Moreover, it appears likely that essentially all of this increase results from the human emission of greenhouse gases." See http://thinkprogress.org/climate/2012/07/28/602151/bombshell-koch-funded-study-finds-global-warming-is-real-on-the-high-end-and-essentially-all-due-to-carbon-pollution/, accessed 28 July 2015.

In biblical terms this is no less dramatic than St Paul moving from Christian persecutor-in-chief to one of the converted after his Damascus experience. One could argue that the tobacco debate was perhaps the one that asked the most dramatic change in societal status quo. Perhaps this is why it took so long to resolve. Or perhaps things just move faster in the digital age. Then again, some majorities are never reached.

88. The term has been used in two prominent books. Walter Lippman, who served with Bernays on the Creel Commission, used it in his book *Public Opinion* (1922). A little over fifty years later came "*Manufacturing Consent: The Political Economy of the Mass Media*" (1988), by Edward S. Herman and Noam Chomsky.

89. See *Munitions of the Mind: A History of Propaganda from the Ancient World to Today*, by Philip M. Taylor, third edition, 2003, online at http://web.elastic.org/~fche/mirrors/www.jya.com/2013/01/aaron-swartz/Mind-Munitions.pdf, accessed 28 July 2015.

90. Later in the nineteenth century, in the Francophone world, Gustave Le Bon wrote extensively on the unconscious and crowd psychology, claiming that crowds needed to be controlled. Another medical scholar, Wilfred Trotter, famous as a surgeon, wrote comparable material in England at the time of World War 1.

91. See http://www.historyisaweapon.com/defcon1/bernprop.html, accessed 28 July 2015.

92. As an example of how propaganda has been used in debates on climate change, let us look at the Heartland Foundation and the controversy that recently erupted over the billboards it erected (one of which is pictured below).

This is admittedly an extreme example – but one that serves to make the point very eloquently.

Before we examine the actual campaign, an introduction to the Heartland Institute can be found on its website at https://www.heartlandfoundation.org/about/vmpv (accessed 23 May 2012) where it is stated: "The Heartland Foundation is a 501(c)3 public charity dedicated to empowering children and adults to build healthier, more livable communities." Elsewhere it describes its core values as inter alia: "Integrity, Accountability, and Ethical Conduct."

The Heartland Foundation does not believe climate change is real. In order to underline this point of view and apparently in an attempt to convince the general public that this is the truth, it devised what must

Photo: The Guardian *at http://www.guardian.co.uk/environment/blog/2012/may/04/heartland-institute-global-warming-murder*

have been a costly media campaign (had it been allowed to run its course). The Heartland Foundation chose four criminals plus Fidel Castro and put their faces on billboards with the words, prominently displayed: "I still believe in Global Warming – do you?"

The group selected to convey this propaganda on the billboards included Osama Bin Laden (who needs no introduction); Ted Kaczynski, who killed two people and injured 23 in a 20-year post-bombing campaign; and Charles Manson (a killer who needs no introduction).

The Heartland Foundation provided its website address at the bottom of the billboards, which were placed alongside very prominent motorist routes in Chicago. In its explanation of the campaign, it stated:

"These rogues and villains were chosen because they made public statements about how man-made global warming is a crisis and how mankind must take immediate and drastic actions to stop it. Why did Heartland choose to feature these people on its billboards? Because what these murderers and madmen have said differs very little from what spokespersons for the United Nations, journalists for the "mainstream" media, and liberal politicians say about global warming. The point is that believing in global warming is not "mainstream", smart, or sophisticated. In fact, it is just the opposite of those things. Still believing in man-made global warming – after all the scientific discoveries and revelations that point against this theory – is more than a little nutty. In fact, some really crazy people use it to justify immoral and frightening behaviour ... The people who still believe in man-made global warming are mostly on the radical fringe of society. This is why the most prominent advocates of global warming aren't scientists. They are murderers, tyrants, and madmen ... Of course, not all global warming alarmists are murderers or tyrants". See http://www.guardian.co.uk/environment/blog/2012/may/04/heartland-institute-global-warming-murder, accessed 28 July 2015.

However, the campaign seems to have misjudged quite badly and has had a strong boomerang effect, leading to some sponsors deserting the Heartland Foundation. However, there are a great many other media and PR campaigns that are more subtle in their disinformation, more insidious and more successful.

The principle of how propaganda works is demonstrated very well: Begin at the end, with the effect that is being induced. Massage the message, package it in the best way possible, don't worry about facts or science more than what will look legitimate, then splash it where it will be as visible as possible and will have the greatest effect possible. Spare no expense.

The contrast with the scientific approach of the IPCC couldn't be greater, nor the mismatch.

93. The sell-by date for solutions to climate change is not very far away. Perhaps ten years, perhaps less. And so it is important to assess the state of play. How close is persuasion and how strong are the forces resisting change? Who is winning?

To understand the way science approaches what is perhaps the greatest challenge of our time, let's use a military situation as an example of what functions have to carried out with success to meet this challenge. In a war, intelligence gathering is most important. Critical questions include: Where will the enemy strike next? What new weapons are they developing? What is the level of their morale? Intelligence is gathered by spies, communications interception and reconnaissance. This is mostly raw data that gets fed back to analysts who interpret the data and decide what it means. For instance, there is a massive build-up of troops and tanks in location A, where "we" are very strong. That could mean "they" intend to attack us where we are strongest, or it could be a diversion (they would never be so stupid, they are trying to draw our attention away from what they are actually planning somewhere else). The analysts are typically highly intelligent people chosen for their skill and experience. Their interpretation then gets fed to the generals who make the decisions.

The men and women making these very important decisions would typically have had extensive training at military academies, where they would have studied the history of war, dissected great battles, and learnt from the thousands of years of case studies. They would also have decades of personal, military experience to inform their decision making. These people are not experts at economics or social development or housing for the poor, or any number of aspects that a politician is expected to deal with in peaceful times, but because the war effort takes precedence over everything else and they know war, they are elevated to very important decision-making positions during wartime.

Now let's compare this to climate change.

Intelligence gathering is done by the Intergovernmental Panel on Climate Change (IPCC). The IPCC is the body of scientists that monitors and assesses state of our global climate. They ask questions and find answers on issues such as: What is happening to global surface temperatures? What is happening to glaciers – are they melting and how quickly? What about rainfall patterns, are they changing or have they become more variable? What is being observed in terms of biodiversity? How is this impacting on food production? What countries are most vulnerable to change and what is their respective ability to adapt? How is this likely to unfold in the future?

The budget of the IPCC is about 6 million Swiss Francs (CHF 6 million) per year. This at time of writing equates to just over US$6 million. As we will see below, this looks rather paltry when one considers that just one oil/chemical conglomerate that regularly funds climate denial has a turnover of about US$100 billion per annum and has spent US$38 million dollars in three years just on lobbying energy related issues. Thus, just one oil/chemical company has spent more in lobbying energy issues than the IPCC spends on hiring offices, paying people, flying them around, writing reports and generally being the custodian of the greatest challenge faced by humankind. Indeed, if somehow there were no overhead costs and all the volunteers working for the IPCC had to be paid equally from the annual budget, they would make about US$2000 to US$3000 per year for the hours and hours and hours of intensive work they do. As it is, they make absolutely nothing.

Yes – being part of the IPCC is voluntary. There is a Secretariat consisting of about a dozen staff – these are paid employees tasked with running the machine. The IPCC publishes what is called an *Assessment Report* every 5–7 years. This is written by about 800 lead authors from all over the world, most often leaders in their respective fields. These scientists are volunteers who do not get paid for their work.

To compare the activities of the IPCC to the military example, this is the intelligence. The data is not raw –the inherent ability of the authors ensures that the reports are focused and relevant.

At this point the baton gets passed to the Working Groups. In a military force, they would be the analysts. The Working Groups combine all the reports into three primary publications. This is done after extensive peer review, with sceptics also being given the chance to comment. In this process the authors have to respond to each comment made. These comments and responses are published. The synthesis report progresses from the "zero order draft" to a final draft. The early drafts are not made available to the public so as to avoid the publication of material that may later change or be deleted (or may be contentious and amended). Only when all the comments have been dealt with are the reports finalised and published.

Perhaps the most important thing the Working Groups do is to create the "Executive Summaries" and "Summaries for Policymakers". The reason why this is so critical is because each of the three reports may be as much as 1,000 pages long and just about nobody reads them. The full documents may be the most important documents published in our time, but it is safe to assume that anybody with the time and enthusiasm to read all of them doesn't have the clout to implement any of the policy imperatives that emanate from the contents.

The "Summaries for Policy Makers" are about 20 pages long and contain virtually everything upon which decisions are made. These documents are agreed upon between the Working Groups and the politicians in an exhaustive, line by line dialogue where the scientists state the science and the politicians try to make it politically correct and to water it down somewhat. This process takes about a week.

The politicians are the generals of the military example and make the decisions that activate the findings. Unlike the generals, they (as a broader group) may not have any specific training for dealing with climate change related issues. The generals often attended military academies, excelled in their studies of past case studies and then completed years of personal experience in war-related situations. As against this, the politicians are most often elected to their jobs by their political parties. These political parties have won elections for any number of reasons and then decided to deploy the specific politician in the field of environmental matters generally (and climate change specifically). Again, for any number of reasons from political reward (because it is seen as a coveted ministry) or political punishment (because it seen as irrelevant) – or perhaps because s/he knows more about that subject than any of the other ministers (not necessarily a great deal). And because we don't think we're in a war, we don't elevate decision-makers on climate change in the pecking order. They still have to deal with all the things generals don't worry about during a war: the economy, social development, job creation, housing for the poor – and getting re-elected.

At this point we come to the critical question of how the IPCC disseminates its findings, puts together marketing campaigns to promote it and engages PR companies on how best to get the message across. Surely, with a message that is critical to the quality of the future, every citizen of every country will have, this is what they do?

Surprise, surprise. The answer is that for the greater part, the IPCC does none of that. The IPCC sees its clients as the member states. Once they have given the "Working Group Reports" and the "Summaries for Policymakers" to the politicians, they step away. Like the analysts in a war say to the generals, they say to the politicians: "There is the data. Do with it as you think best."

They see their mandate as being neutral. Being policy relevant but not prescriptive. They will give choices and scenarios but for the greater part they believe that there is no need for them to popularise what they are doing. They think this is the task of governments and international bodies like the United Nations Environment Programme (UNEP). Indeed, for a long time the IPCC did not have a media liaison.

In summary then, the game plan for the science team is to work preferably from peer reviewed research, to use hundreds of volunteer scientists (who have other day jobs and are leaders in their fields) to contribute reports, to compile everything into three primary reports, to produce "Executive Summaries" and then to step away and let the politicians take over.

94. It was said of Winston Churchill that ..."He knew people would not march behind a subordinate clause." See http://www.nytimes.com/2001/11/11/books/his-finest-hour.html?pagewanted=all, accessed 11 October 2012. And Trotter says, somewhat cynically:

"It has already been pointed out how dangerous it would be to breed man for reason – that is, against suggestibility. The idea is a fit companion for the device of breeding against 'degeneracy'. The degenerate – that is, the mentally unstable – have demonstrated by the mere fact of instability that they possess the quality of sensitiveness to feeling and to experience, for it is this which has prevented them from applying the remedy of rationalization or exclusion when they have met with experience conflicting with the herd suggestion, ... it is interesting to notice that in discussing the mechanism of psychoanalysis in liberating the 'abnormal' patient from his symptoms, Freud repeatedly lays stress on the fact that the efficient factor in the process is not the actual introduction of the suppressed experiences into the conscious field, but the overcoming of the resistances to such an endeavour. I have attempted to show that these resistances or counter-impulses are of environmental origin, and owe their strength to the specific sensitiveness of the gregarious mind. Resistances of similar type and identical origin are responsible for the formation of the so-called normal type of mind. It is a principal thesis of an earlier essay in this book that this normal type is far from being psychologically healthy, is far from rendering available the full capacity of the mind for foresight and progress, and being in exclusive command of directing power in the world, is a danger to civilization." Excerpt from W. Trotter, *Instincts of the Herd in Peace and War* (Unwin, London, 1919), pp 63–64.

PART 4 BEAUTIFUL TRUTHS

Chapter 26: The back door of perception

95. See http://www.etymonline.com/index.php?term=physicist, accessed 30 July 2015.

96. "... my endeavours should be directed to persons and characters supernatural, or at least romantic, yet so as to transfer from our inward nature a human interest and a semblance of truth sufficient to procure for these shadows of imagination that willing suspension of disbelief for the moment, which constitutes poetic faith ..." (emphasis added). See http://goo.gl/H9xVdO, p 2, accessed 11 October 2012.

97. See J.R.R. Tolkien's compelling essay "On Fairy-Stories" at http://bjorn.kiev.ua/librae/Tolkien/Tolkien_On_Fairy_Stories.htm, accessed 11 October 2012. Here the need for inherent consistency in fiction is discussed in perhaps the most challenging genre from this perspective, being fairy tales. Tolkien writes:

    > Fantasy has also an essential drawback: it is difficult to achieve. Fantasy may be, as I think, not less but more sub-creative; but at any rate it is found in practice that 'the inner consistency of reality' is more difficult to produce, the more unlike are the images and the rearrangements of primary material to the actual arrangements of the Primary World. It is easier to produce this kind of 'reality' with more 'sober' material. Fantasy thus, too

> often, remains undeveloped; it is and has been used frivolously, or only half-seriously, or merely for decoration: it remains merely 'fanciful'. Anyone inheriting the fantastic device of human language can say the green sun. Many can then imagine or picture it. But that is not enough –though it may already be a more potent thing than many a 'thumbnail sketch' or 'transcript of life' that receives literary praise ....To make a Secondary World inside which the green sun will be credible, commanding Secondary Belief, will probably require labour and thought, and will certainly demand a special skill, a kind of elvish craft. Few attempt such difficult tasks. But when they are attempted and in any degree accomplished then we have a rare achievement of Art: indeed narrative art, story-making in its primary and most potent mode.

98. See the discussion by psycho-analytical literary critic, Norman N. Holland, *The Willing Suspension of Disbelief: A Neuro-Psychoanalytic View* available online at http://www.psyartjournal.com/article/show/n_holland-the_willing_suspension_of_disbelief_a_ne. Holland investigates the reasons and mechanics behind these phenomena. With references to an episode where he went to see "Love Story" in a theatre frequented by "rednecks and bikers" (the heroine dies of leukaemia) he posits that:

> In short, we can feel real emotions toward unreal fictions, because two different brain systems are at work. One, the prefrontal cortex's inhibiting system is at work because we know we are not supposed to act in response to the fiction we are reading or the drama we are watching. We therefore cease to test reality and we do not disbelieve the fiction. But our corticolimbic system remains at work, and through it we feel the emotions we would ordinarily feel at the human situations we are watching, at, say, Jenny Cavalieri's love and death ... In the case of the willing suspension of disbelief, the prefrontal cortex inhibits action and the planning of actions so that we no longer are aware of the unreality of the fictions we are dealing with, but it does not – cannot – inhibit the corticolimbic systems that give rise to our emotions. They run freely on, busily prompting us to actions, to approaches and avoidances, we never perform, but the

psychological feelings and the physical signs of emotion persist. I cry and my fellow-spectators, the bikers and rednecks, cry for Jenny Cavalieri.

99. Jonathan Gottscall, a pioneer in the field of literature and evolution, says the following:

> ... research consistently shows that fiction does mould us. The more deeply we are cast under a story's spell, the more potent its influence. In fact, fiction seems to be more effective at changing beliefs than non-fiction, which is designed to persuade through argument and evidence. Studies show that when we read nonfiction, we read with our shields up. We are critical and skeptical. But when we are absorbed in a story, we drop our intellectual guard. We are moved emotionally, and this seems to make us rubbery and easy to shape (emphasis added).

He says further, that fiction creates a greater ability to empathise:

"But perhaps the most impressive finding is just how fiction shapes us: mainly for the better, not for the worse. Fiction enhances our ability to understand other people; it promotes a deep morality that cuts across religious and political creeds. More peculiarly, fiction's happy endings seem to warp our sense of reality. They make us believe in a lie: that the world is more just than it actually is. But believing that lie has important effects for society – and it may even help explain why humans tell stories in the first place". See generally Jonathan Gottschall, *The Storytelling Animal: How Stories Make us Human* Houghton Mifflin Hardcourt 2012); and specifically, "Why Fiction is Good for you", *Boston Globe*, 29 April 2012, at https://www.bostonglobe.com/ideas/2012/04/28/why-fiction-good-for-you-how-fiction-changes-your-world/nubDy1P3viDj2PuwGwb3KO/story.html, accessed 28 July 2015.

An investigation of the scholarly works that Gottschall cites in his discussion is illuminating:

- J. González, A. Barros-Loscertales, F. Pulvermüller, V. Meseguer, A. Sanjuán, V. Belloch and C. Ávila, "Reading Cinnamon Activates the

Olfactory Regions of the Brain", *NeuroImage*, Vol. 32, No. 2, 2006, pp 906–912. The implication is that reading words strongly associated with smell activates the corresponding part of the brain in the primary olfactory cortex.

- R.A. Mar, K. Oatley, J. Hirsh, J. de la Paz, and J.B. Peterson, "Bookworms versus Nerds: Exposure to Fiction versus Non-fiction, Divergent Associations with Social Ability and the Simulation of Fictional Social Worlds", *Journal of Research in Personality*, Vol. 40, 2006, pp 694–712. Reading fiction correlates with improved social interaction and empathy, while reading non-fiction does not.
- See V. Boulenger, O Hauk and F. Pulvermüller, "Grasping Ideas with the Motor System: Semantic Somototopy in Idiom Comprehension", *Cerebral Cortex*, Vol. 19, 2009, pp 1905–1914. The brain "primes" body parts referred to in idiomatic expression more pronouncedly than in literal sentences.
- R.A. Mar, K. Oatley and J.B. Peterson, "Exploring the Link between Reading Fiction and Empathy: Ruling out Individual Differences and Examining Outcomes", *Communications*, Vol. 34, 2009, pp 407–428. Exposure to fiction enhances empathy and positively correlates with social support, while exposure to non-fiction is positively correlated with loneliness.
- R.A. Mar, "The Neural Bases of Social Cognition and Story Comprehension", *Annual Review of Psychology*, Vol. 62, 2011, pp 103–134. Mar posits that the "mentalising" network that is used to understand the state of mind of others, overlaps with the narrative comprehension network. This conclusion suggests the value of the narrative in eliciting empathy.

100. Ziva Kunda "Social Cognition – making sense of people" MIT Press 5th edition 2002 p 101 makes the point that research shows that if people imagine an event like winning the lotto, they will afterwards deem it more likely to occur than those who had not made the effort to imagine it. By the same logic, imagining what environmentalists are saying about PECCAs to be true would in all probability cause the person doing the imagining to be more likely to believe it to be true. On p 137 – 139 Kunda relates research showing that members of a jury spontaneously construct stories to tie together pieces of evidence.

Chapter 28: Beautiful truths

101. Nietzsche The Birth of Tragedy Vintage Books a division of Random House, translated by Walter Kaufmann 1967 chapter 14 p 93, says: "It was something akin to the demonic warning voice that urged him to these practices; it was his Apollonian insight that, like a barbaric king, he did not understand a noble image of a god and was in danger of sinning against a deity—through his lack of understanding. The voice of the Socratic dream vision is the only sign of any misgivings about the limits of logic. Perhaps – thus he must have asked himself – what is not intelligible to me is not necessarily unintelligent? Perhaps there is a realm of wisdom from which the logician is exiled? Perhaps art is even a necessary correlative of and supplement for science?"

102. I am indebted to Professor Emily Wilson, author of "*The Death of Socrates*" for kindly and quickly providing the reference to the Plato's Phaedo to me (a complete stranger) at a time when I had lost it and could no longer find the passage that had led to the title of this chapter some months before, when I did my planning. It is the Jowett translation.

103. WB Yeats in "The Collected Works of W.B. Yeats in "Volume XIII: A Vision" said: "Yet even the truth into which Plato dies is a form of death, for when he separates the Eternal Ideas from nature and shows them self-sustained he prepares for the Christian desert and the Stoic suicide." – see https://books.google.co.za/books?id=QxVvh5BHkloC&pg=PA43&lpg=PA43&dq=he+prepares+for+the+Christian+desert+and+the+Stoic+suicide.".

About the author

Johan van den Berg is a barrister and renewable energy practitioner. He is the CEO of the South African Wind Energy Association, the Chair of the South African Energy Council, and the African Private Sector Focal Point for the Africa–EU Energy Partnership.

He serves as a member of the South African Ministerial Advisory Commission on Energy and was awarded the 2015 Global Sustainable Leadership Award by the World CSR congress in Mumbai, India. He lives in Johannesburg, South Africa.